FANTASTIC PLASTIC

THE KITSCH COLLECTOR'S GUIDE

FANTASTIC PLASTIC

THE KITSCH COLLECTOR'S GUIDE

PETE WARD

CHARTWELL
BOOKS, INC.

A QUINTET BOOK

Published by Chartwell Books
A Division of Book Sales Inc.
114 Northfield Avenue
Edison, New Jersey 08837

This edition produced for sale in the U.S.A., its
territories and dependencies only.

ISBN 0-7858-0734-9

This book was designed and produced by
Quintet Publishing Limited
6 Blundell Street
London N7 9BH

Creative Director: Richard Dewing
Art Director: Clare Reynolds
Designer: Melvyn Walker
Project Editor: Kathy Steer
Editors: John Wright and Sean Connolly
Text Contributor: Percy Reboul
Photographer: Jeremy Thomas
Additional photographs: John Morgan

Typeset in Great Britain by
Central Southern Typesetters, Eastbourne
Manufactured in Singapore by
United Graphics Pte Ltd.
Printed in China by Leefung-Asco Printers Ltd.

CONTENTS

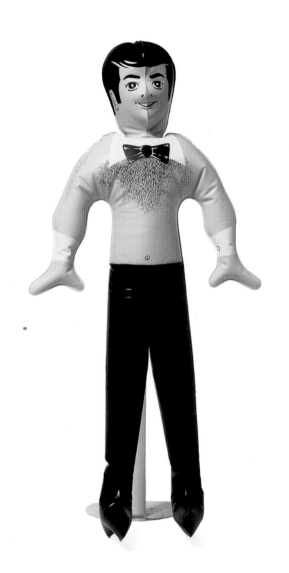

Despite the armies of style gurus, correspondents, and experts who are constantly telling us on television, in weekend supplements, and countless glossy magazines what to wear and eat or how to decorate our homes, bad taste will not go away. It clings like a super virus, multiplying endlessly, resistant to all known aesthetic detergents, nourishing itself on the seemingly endless supply of products of our popular culture.

Think of bad taste as a kind of warehouse for fashions whose time has run out, decor that has disappeared, novelties that make you wince, and celebrities past their best-before date. When this merchandise has been lying around gathering dust for long enough, it appears to acquire a new aura. The previously unmitigated hideousness starts to suddenly shine with a new appeal and is recharged with new energy. What was previously just awfulness starts to look like naughty fun and we all want some. Suddenly we miss them and remember fondly those long-gone eras of exuberance. It has happened to objects as diverse as Pez dispensers, Liberace, and snow globes.

Now you can understand just how drab, dull, and uninteresting our world would be without kitsch. It is our savior from the stultifying predictability of the taste-mongers of the shopping malls and out of town discount stores that dominate the retail landscape.

Free the mind, learn to love kitsch.

KEY TO PRICE GUIDE

A price guide appears above every caption throughout this book:

✪	*Do not part with any more than a few dollars*
✪✪	*Has a value of between $10-$20*
✪✪✪	*Worth in the range of $20-$50*
✪✪✪✪	*Highly prized collectible or rare; pay up to $100*
✪✪✪✪✪	*An absolute gem of an artifact that could cost you anything above $100*

W hen the prolific British inventor Alexander Parkes (1813-1890) decided to show the general public for the first time a range of products made from his new material Parkesine, he shrewdly chose as the venue the International Exhibition held in London in 1862. Parkesine was the world's first plastic material – a form of nitrated cellulose which was to become better known under the name Celluloid. Neither Parkes nor any of the six million people who visited that exhibition could have realized that here was the beginning of a great industry supplying the world with an essential raw material.

The exhibition catalog recorded Parkes's claims for his product: "a material as hard as horn, flexible as leather, capable of being stamped, painted, dyed and even carved and which, above all, can be produced in any quantity at a price lower than gutta percha." He was wrong in his assessment of the manufacturing costs and it was not long before Parkes faded from the picture. He had, however, sown the seeds of plastics invention and was to be followed in succeeding generations by a series of great inventors and scientists who discovered and developed plastics such as Bakelite, polyethylene, PVC, nylon, and a host more.

But Parkesine and the plastics that followed it did far more than merely substitute for natural raw materials. Because plastics can be molded or otherwise shaped by a repetitive process, plastic items could be mass-produced and thus meet an increasing demand from both industry and people who, for the first time, had a little money to spend on luxuries. An early example of this was the invention of the washable Celluloid collar and cuff which made a modest fortune for the early plastics industry. Now the emerging clerical classes of Victorian England could emulate the bosses by wearing a clean collar every day!

It is interesting to reflect that among the early uses of Parkesine were personal adornment articles such as earrings, necklaces, bracelets, and the like. These were to be followed by cutlery handles and combs but we may reflect here that, because Celluloid could be made into very good substitutes for ivory and tortoise shell, Parkes and his successors certainly saved the elephant and turtle from extinction! Were these products, one wonders, regarded as kitsch? Certainly they were nothing like the quality of those products made individually by craftsmen from natural materials.

Plastics have become indispensable to modern living, not least because of their immense versatility. Plastic items are not only produced in almost any shape or color but they can be formulated to give specific technical properties such as toughness, corrosion resistance, electrical and thermal insulation, flexibility, rigidity, and so on. Thus we have, at one and the same time, a raw material upon which the world depends but which is also used to produce hideous, outrageous trivia regarded by most people as a waste of the earth's limited resources. Somewhere within this spectrum lies the fascinating world of kitsch whose delightful eccentricities appeal to so many collectors. Which is where this much-needed book comes in. I, for one, will be looking forward with keen anticipation to meeting within its pages friends old and new and getting confirmation that there are others out there who share my great enthusiasm for collecting such strange objects.

Percy Reboul
Plastics Historical Society
May 1997

PLASTIC EXTRAVAGANZA

1

1

Kitsch and plastic seem made for each other, a marriage made in heaven. Plastic is a generic name for a range of materials developed to impersonate increasingly rare and expensive natural materials, and "kitsch" is a word with its origins in the Austrian vernacular expression "verkitschen etwas," meaning to knock off (produce) something that is sickly sentimental.

Fundamental to understanding what kitsch is all about is the notion of taste itself. It was the outrageous film director John Waters (*Hairspray, Cry-Baby, Serial Mom*) who said: "In order to acquire bad taste, one must first have very, very good taste."

Taste is the faculty we all have for enjoying and discerning beauty. Like any exercise of our judgment, decisions as to what is and what is not tasteful are based upon an amalgam of influences from different times and different places. The fact that we all have widely differing likes and dislikes is as old as the hills. There is even a Latin saying, "de gustibus non est disputandum," meaning there is no dispute about taste. This is the earliest formulation of the old chestnut that "one man's meat is another man's poison," confirming the eighteenth-century philosopher David Hume's proposition that "beauty is not a quality that is inherent in things, but exists in the eye of the beholder," and contradicting the philosophy of his contemporary Immanuel Kant who argued that aesthetic judgments have universal validity.

Neither of these two great minds could anticipate the explosive effect that the imminent Industrial Revolution was going to have on the issue of taste. The era of mass production had the effect of turning all the social classes into consumers and, in turn, threw the question of taste away from the social classes of the eighteenth-century tea-rooms and onto the streets.

By the start of the Victorian era, there had been a sharp increase in demand for goods and merchandise, fuelled by a vast expansion in the supply capacity due to technological advance. A new industrial culture based upon "things" had emerged.

Many high-minded people found much of the produce of this new mass culture distasteful. So

✪

Left: Holiday souvenir stalls, kiosks, and stores around the world are a fine place in which to find those collectible kitsch mementoes. The very best are nearly always rendered in a plastic of one sort or another, and this is no exception. A souvenir from the quaint British holiday island, the Isle of Wight, containing different colored sands that the isle is famous for.

✪ ✪ ✪

Above: This '70s globular alarm clock reflects many of the obsessions that designers had at the time: futuristic objects, aerodynamic designs, and the space race, plus the color pink. This example is an original and is authenticated by the yellowing of the originally white plastic casing.

✪ ✪ ✪

Center: The pink flamingo is a bit of a kitsch icon, deified by the filmmaker John Waters in his low-budget 1972 comedy movie, *Pink Flamingos*, that launched him to superstardom. These colorful wonders graced nearly every garden lawn and pond in the '50s and '60s. Nowadays of course, it's all beechwood and oak furniture and natural colors. Put a little sparkle back into your garden. Combined here with a plastic '50s vase and some plastic '50s chysanthemums, here's a display that works equally well both indoors and outdoors.

✪ ✪

Above: An acrylic jam or marmalade pot set with stunning floral display within the lid. This object dates from the late '70s but is still currently manufactured. Its one advantage over conventional confiture carrying devices is that, being plastic, it can sustain a drop onto the kitchen floor and still contain the contents.

13

concerned were they that in the U.K., Prince Albert, no less, together with a few friends, set out to instil a few aesthetic principles into the middle classes with the Great Exhibition of 1851. Every exhibit was chosen "for its merits in exemplifying some right principle of construction or ornament," according to the self-righteous publicity.

However, this attempt to aim the bourgeoisie in the right direction ultimately failed, for as soon as there was a demand for objects which signaled to the world the message "I have good taste," and these objects were able to be mass-produced, then consequently anybody could possess them and claim the status of good taste. This bandwagoning by the bourgeoisie, far from heralding the dawn of a glorious new era in universal discrimination, was to mark the origin of the wonderful perversion of taste that we know as kitsch.

The word itself came into use in the early part of the twentieth century, used by professional art critics to describe the worst manifestations of mass taste. In 1925, an Austrian art critic, Fritz Karpen, entitled a pamphlet "Der Kitsch." This was a treatise devoted to criticizing contemporary designs of the times that appeared crudely inappropriate to their purpose, such as a poker ironwork of Da Vinci's The Last Supper and breasts as inkwells.

The word "kitsch" had extra dismissive power because it derived from a slang expression, adding extra contempt to the feeling these art critics had for these objects. Some critics perceived kitsch as a far more insidious and far-reaching monster. In 1933, Herman Broch commented that "all periods in which values decline are kitsch periods. The last days of the Roman Empire produced kitsch and the present period (the era of the Third Reich) . . . cannot but be represented by aesthetic evil."

Some of these critics were almost visionary in their predictions. In 1939, Clement Greenberg warned that the spread of bad taste was "a virulence of kitsch" that had gone on a triumphal world tour, so that "by now it is becoming the first universal culture ever." For Greenberg, it was the language through which the majority of people made sense of their lives. With the steady growth of popular culture since then, this analysis would mean that the entire century has been an endless spiral of cultural decline.

Greenberg's nightmare has inflated beyond his worst imaginings. Today, the planet is overflowing with bad

✪ ✪ ✪ ✪

Below: This almost unbelievable piece of plastic fish-shaped basket-work is in fact a handbag. It is mid '50s and a really comical and very rare example. Wear matching pink coat and shoes, and pop on the sunglasses to complete the effect.

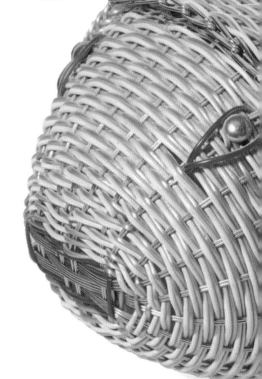

✪ ✪

Left: Another variation on the handbag theme. This glorious item is in reality a portable transistor radio from the mid '60s with the extra value of a message of love embossed on the front. It makes an ideal gift for the lady in your life for Valentine's Day.

✪ ✪

Right: This jolly sailor snow globe has such a large appetite he can eat an entire English county, as displayed in his transparent stomach. This globe dates from the late '60s.

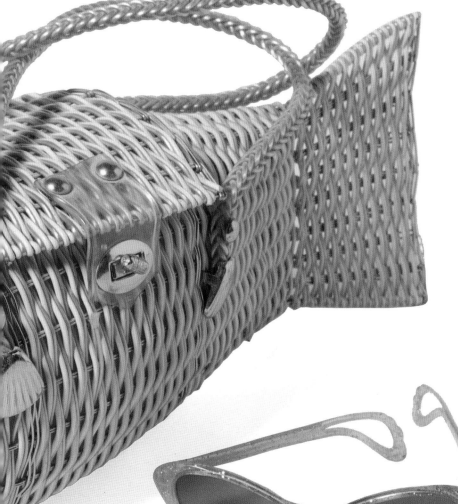

✪ ✪

Right: Here is another snowglobe. This time, the fisherman has caught a huge fish in a very fierce snow storm.

taste, from plastic donkeys dispensing cigarettes from their rears to nylon/acrylic Elvis Presley rugs.

Kitsch is a complex subject. It is still defined in dictionaries as "worthless pretentiousness in the arts," which simplistically aligns it with bad art, or as "tawdry vulgarized popular art usually with sentimental appeal" which identifies it with even worse art. Neither definition gives you much idea of what kitsch really is nor recognizes quite how popular it has become. On the other hand, the current colloquial usage is just too loose to be useful since the word is often applied to anything that could be regarded as simply bad taste, ranging from the outrageously offensive to high camp.

Kitsch is not simply bad taste, even though it is a creature of bad taste. It has the powerful ability both to attract and repel at the same time. It holds an almost childlike fascination for us. This is hardly surprising given the gaudiness of the colors, the unusual shapes, the lack of sophistication, and the nostalgic associations that kitsch has. As it repels us by the way it clashes with our notions of good taste, within this very awfulness lies its real appeal. It was the poet Charles Baudelaire who summed up this essential kitsch experience best: "What is so intoxicating about bad taste is the aristocratic pleasure of being displeased." A feeling of superiority creeps up on us derived from the fact that there are people out there, who, without any sense of irony, genuinely draw comfort and pleasure from tasteless objects.

In determining the presence of kitsch, it is essential to detect a few vital clues. Just as refinement and dignity are integral to good taste, so excess and impropriety are the innate properties of kitsch. There is an inane and comic crudity present in a set of pastel-colored drinking straws depicting copulating couples molded into the stems. If there is a sense of value attached to an object considered to be in good taste, the kitsch object will devalue

✪ ✪ ✪

Above and below: Snow globes form the cornerstone in many a kitsch collector's collection. The most common, of course, are the snowscapes from the world's famous cities and holiday destinations. Here are very witty and unusual variations on the theme. The smog is a mixture of grit and dust is suspended in the water instead of plastic shavings (original '30s snow globes contained bone chips). Pollution is, after all, one of the things Los Angeles has been remembered for, as much as Hollywood.

✪ ✪

Left: This '60s go-go dancing girl is actually a cocktail shaker. Just add batteries, press the button, and off she goes. A marvellous margarita in 10 fun seconds!

this notion (e.g. a Mona Lisa keyring). A sense of inappropriateness is important too, as in stubbing out your cigarette in a plastic ashtray souvenir of the Vatican City.

In summary, kitsch simply wouldn't exist at all were it not for the public's endless and fickle quest for owning "things." Propelled by the commercial pressures of advertising, the media, and social status, our society produces a vast amount of objects which are sought after, fought over, acquired, and finally discarded as public taste moves on in search of new ideas and new markets. Without the hideous wasteful aspect of our economic system, kitsch would not exist.

✪✪

Above: Furry dice are an instantly recognizable all-time kitsch classic. A mobile kitsch statement. No automobile is complete without a set. Made from a sort of brushed nylon/acrylic-like material. Originals from the '60s are very hard to find and have usually discolored or grown fungus due to the elements. However, new models such as these are now being manufactured again in the Far East.

✪✪✪

Above: Not only a nodding dog but a poodle as well. Two great kitsch icons rolled into one. Original '50s dogs fetch a lot of money but are rarely in good condition. Pristine new versions are now being manufactured for the fast-growing kitsch-conscious market. Looks great if combined with the furry dice.

✪

Below: Kitsch doesn't always have to be about the past. This is a modern-day air freshener in the shape of a crown, currently available at gas stations worldwide. A royal word of warning: do not open! The odor is so pungent, it will make your home or car smell like a bathroom.

1

The kitsch lover is a dedicated environmentalist, a cultural recycler sifting through the ever-proliferating stock of schlock in the world and rescuing what makes us laugh, thumbing the nose at the notion of good taste, social rules, and codes.

Kitsch has outgrown its cult status and become a source of inspiration for a great deal of fashion, interior design, and comedy; a serious issue of discourse in the fine arts; and the basis for a thriving industry of second-hand merchandise. Following the laws of supply and demand, there is also a thriving industry producing new goods and entertainments to be appreciated as kitsch, deliberately.

Save the planet! Collect kitsch now!!

✪ ✪

Below: A very amusing smoker's accompaniment – this bizarre donkey cigarette dispenser distributes its bounty from its backside. Unusual and dates back to the '60s. Very collectible.

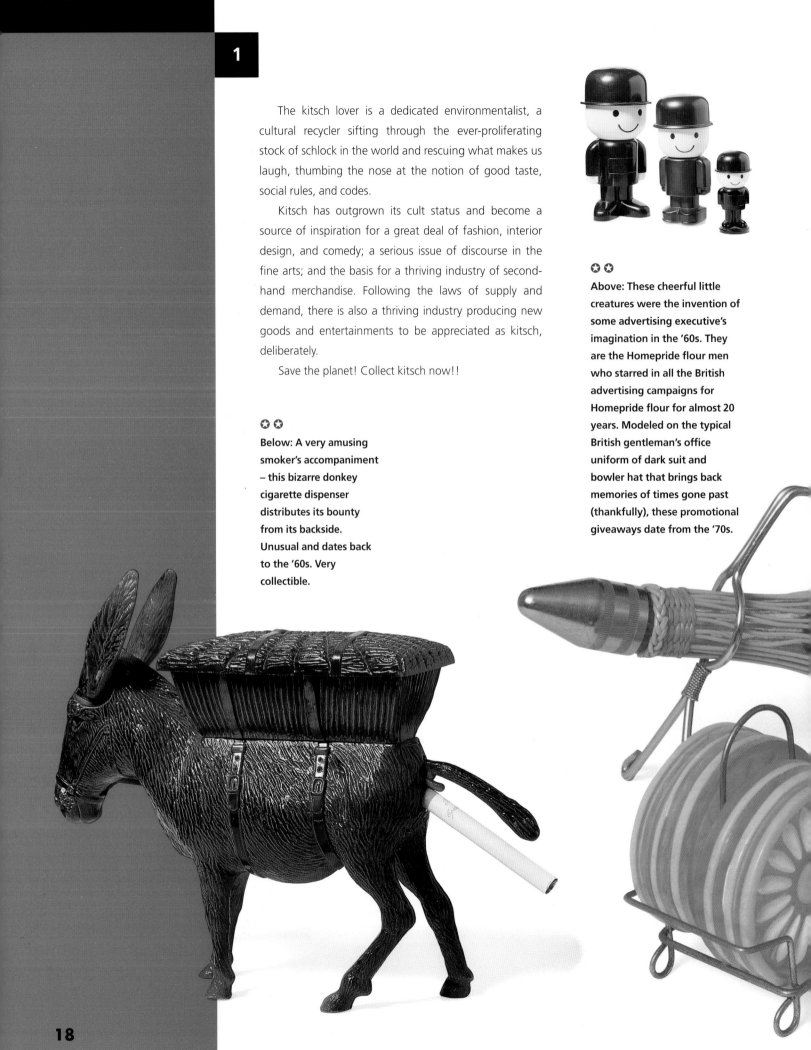

✪ ✪

Above: These cheerful little creatures were the invention of some advertising executive's imagination in the '60s. They are the Homepride flour men who starred in all the British advertising campaigns for Homepride flour for almost 20 years. Modeled on the typical British gentleman's office uniform of dark suit and bowler hat that brings back memories of times gone past (thankfully), these promotional giveaways date from the '70s.

✪ ✪ ✪

Right: This acrylic paperweight is made from a real plastic bag of sweets designed by Paul Clark and is dated to the early '80s. A very effective paperweight, but it also works very well as an objet d'art in itself.

✪ ✪

Left: Like the Homepride men, these doe-eyed deer were adopted as a gimmick, this time for a popular cheap and sweet British champagne substitute, Babycham. These are promotional items from the '70s.

✪ ✪

Below: This beautiful and vivid range of products would bring a little ray of sunshine into any home. The bottle at the back of the display once contained sangria. The floral coasters are heat resistant, and the floral arrangements and fish tanks are from Taiwan. They all date from the '70s.

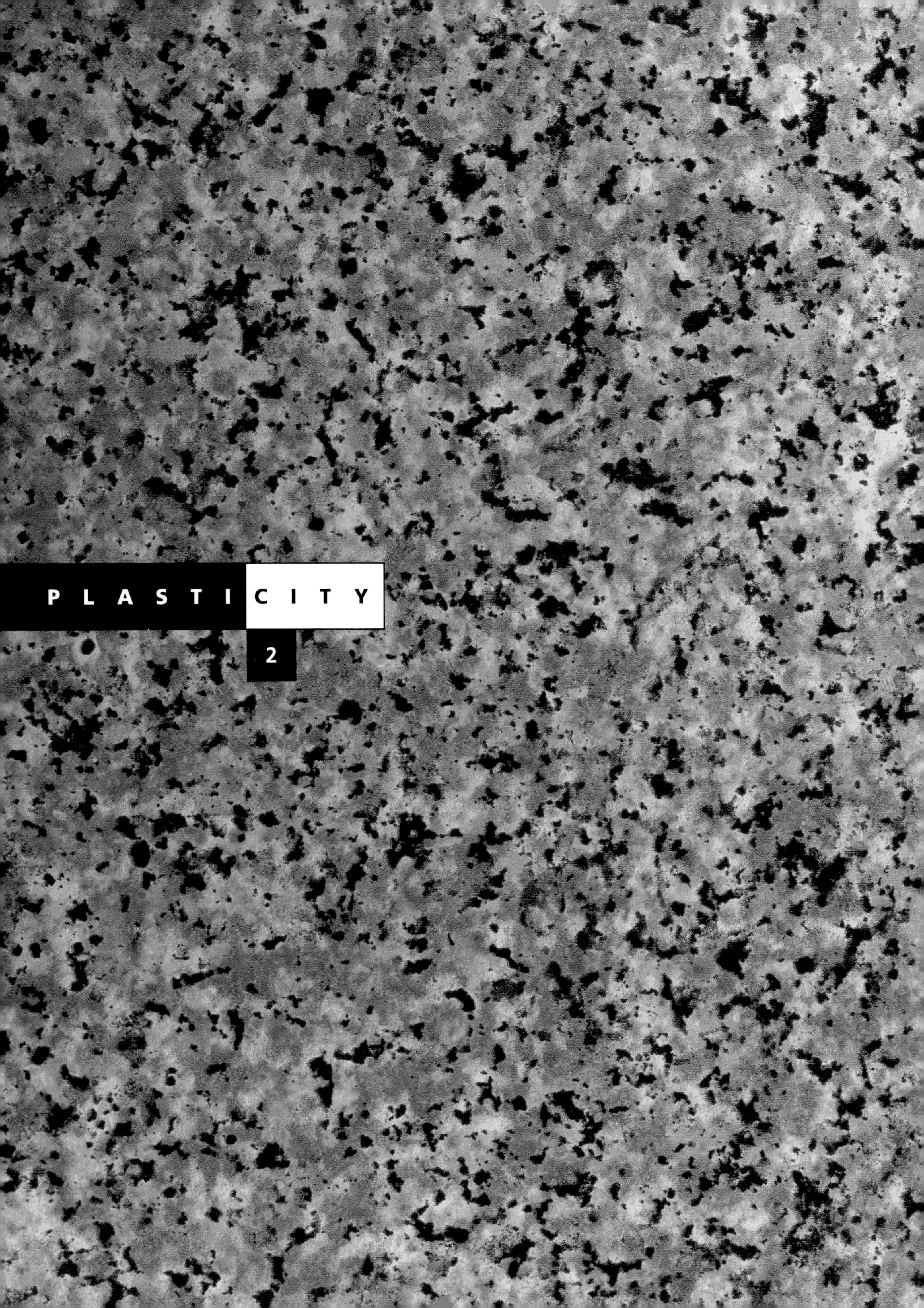

P lastics are largely the products of twentieth-century science and technology, and their large-scale use has quite simply transformed the world in which we live. Plastics are so familiar, we take them for granted and often misunderstand them.

A common belief is that they are all variations of one material called "plastic." In fact, they comprise a whole group of materials, each one chemically different and with its own properties. Some plastics, for example, have excellent chemical resistance, others are good insulators or may be foamed to provide lightweight shock-resistant packaging materials. Yet another important group can be produced in the form of fibers for use in the textile industry.

Plastics are not the ultimate solution to all our technical problems, but, on their own or in conjunction with other materials such as metals and ceramics, they offer designers and engineers a huge range of properties with which to work.

A common misconception is that plastics are simply cheap substitutes; they are in fact often more expensive than more traditional materials. Few, however, can match the cost-effectiveness of plastics brought about by the lightness in weight, ability to be mass-produced, wide range of properties, and consistent performance in service.

SO WHAT ARE PLASTICS?

Curiously, it is not easy to define plastics in simple terms. This is because, like wood and metal, they are a large family of materials which have certain features in common. For example:

All plastics are **polymers**. Polymers are giant molecules which occur both in nature or can be man-made. A molecule of water, for example, is formed of three atoms (H_2O). A typical plastics molecule can comprise 100,000 atoms. All plastics can be **formed** into shape. Plastics soften when heated; i.e. become plastic, and it is at this stage that they can be molded or otherwise formed into their final shape—usually by heat and pressure.

Most plastics are **organic**, i.e. they have a molecular "backbone" of carbon atoms.

Plastics fall into two types:

Thermoplastics are those which will soften when heated and harden on cooling—a process which can be repeated over and over again. Vinyl and polyethylene are well known examples.

✪ ✪

Above: This selection of domestic goods demonstrates the real practicality of the early plastics; here we have darning shoes, a salt cellar, and a canapé dish. Kitsch wasn't common in those early days and it wasn't until the later polyera of plastic that really frivolous kitschy objects were to see the light of day.

✪ ✪ ✪

Above: A selection of kitchenware items. All the items are made from mottled urea formaldehyde from about mid to late '30s.

Thermosets are those materials which soften and harden only once when they are first heated. Examples are phenolics (Bakelite), melamine, and polyester resins.

Plastics can also be classified in another way as natural, semisynthetic, and synthetic.

THE NATURAL PLASTICS – HORN

In the seventeenth century an Englishman, John Osborne, made moldings from a natural polymer, horn, but for hundreds of years before him substances such as wax, amber, bitumen, and other natural organic polymers had been molded by man. The horn industry today is nothing like as large or prosperous as it once was having become a victim of the commercial success of modern plastics with its speed of production and ready source of raw materials. Items of horn kitsch do exist and it is of interest that the Worshipful Company of Horners, whose origins go back to the thirteenth century, now incorporates people from the plastics industry.

In America, shellac obtained from the lac beetle living on certain trees in India, Burma, and Thailand was being mixed with wood flour to mold Union Cases to house early photographs. Because shellac moldings could incorporate fine detail, it was still being used into the '50s to make 78 rpm gramophone records which were replaced eventually by another plastics, vinyl.

Gums from tropical trees were exploited, especially rubber and gutta percha for which Bewley invented the extruder in 1847. Gutta percha, with its excellent resistance to water and good electric insulation properties, was used to protect the first submarine telegraph cables in 1850. It was also used to make a host of moldings ranging from ear-trumpets to golf balls. Charles and Thomas Hancock, great inventors, worked extensively with this material and Thomas was the first to patent his discovery of the vulcanization of rubber which Goodyear discovered independently in America. Theirs was the first deliberate chemical modification of a natural resin to produce a new molding material.

In 1855, in France, Lepage and Talrich patented a molding compound which they called Bois Durci. It was made from wood flour and albumen obtained from blood and egg whites. One of its most famous applications developed by another Frenchman, Latry, was for portrait plaques. Many others worked with a variety of ingredients including

✪✪✪
Above: Victorian molded horn brooch. Quite rare now.

✪✪✪✪✪
Left: One of the early shellac Union Cases.

✪✪✪✪
Above: An exquisite pipe made out of vulcanite.

seaweed, peat, paper, and leather. In fact, nearly ten percent of all British patents issued in 1855 referred to molding materials but it was to be the breakthrough in the modification of cellulose fiber with nitric acid that was to give the world its first semisynthetic plastics material, cellulose nitrate.

THE EARLY DAYS OF PLASTICS – SEMISYNTHETICS

Celluloid The birth of the plastics industry took place at the International Exhibition held in London in 1862 when a gifted British inventor, Alexander Parkes showed a new material which he called **Parkesine**. It was, in fact, the world's first man-made plastics (a form of nitrated cellulose) but neither Parkes nor any of the six million visitors to the show could surely have imagined that here was the beginning of a vast international industry whose products would transform the world.

Parkes was just one of the 29,000 exhibitors taking part and his exhibits included medallions, knife handles, and combs.

Parkes was no businessman. He failed to make a commercial success of his invention and sold out to another Englishman, Daniel Spill, who made a number of improvements and called his products by the wonderfully futuristic names of **Xylonite** and **Ivoride**. He too, however, over a period of years, suffered a number of setbacks connected with financing and patent infringement from which he never recovered. A decade or so was to pass before L.P. Merriam, an American living in Britain, came on the scene with his version of Xylonite. This was a more successful product but it was to take years before his company, British Xylonite, finally turned the corner with products such as combs, collars, and cuffs.

At the same time that all was going for Spill, an American billiard-ball company offered a $10,000 prize to anyone who could find or manufacture a substitute for ivory, a rapidly diminishing commodity causing an equally swift rising price in the cost of manufacturing billiard balls.

An American inventor, John Wesley Hyatt, accepted the challenge and, although he didn't

Above: These Celluloid, stylized Scottie dogs and cats as brooches and as pins are about as close as you can get to sentimental kitsch with this era of plastic production.

Left: An early organ grinder made out of polystyrene. Quite exquisite and rare.

Below: Celluloid bracelets. The white one has been opalized through the use of ground shell.

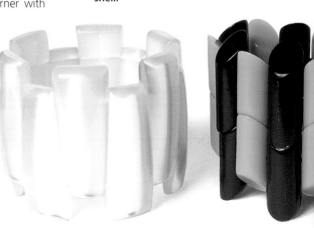

Left: A piece of Celluloid. An ox and cart. Quite rare.

Below: Celluloid cherry necklace and bracelet. All molded from Celluloid, including the chain and leaves.

Below: An excellent example of casein. The birds are sitting aside the ashtray. The base is wooden, quite rare now.

Above: This wacky Bakelite and acetate cat has the most wonderful, googly eyes.

actually win the prize, his experiments yielded a ball with a core of molded gum shellac and pulp, coated with collodion, a solution of cellulose nitrate and alcohol that dries in contact with air to a thin film. Like Parkes's cellulose nitrate, it was also highly flammable and, unfortunately, when the balls hit against each other during a game of billiards, the collodion was said to explode with a mild detonation that sounded like a gunshot. Hence the failure to win the prize.

But the billiard-ball business loss was Hyatt's gain. These problems led him to further experiments with the raw materials and in 1870 he obtained a patent for "Celluloid." This material provided the basis for the business empire he established with his brother: the American Celluloid Company, today the Plastics Division of the Celanese Corporation.

Celluloid was the American trade name for the first commercially successful man-made plastics. It was a great novelty, not to say marvel of the age. It was an entirely new material that could be formed into virtually any shape. It was the first of many new plastics materials before the first half of the twentieth century.

Hyatt also devised the blow-molding process in which a hollow tube of Celluloid held in a special press was heated and then inflated with hot air until the tube conformed to the shape of the surrounding mold. The result was a three dimensional hollow object of great delicacy. The principle is still in use today.

While Celluloid was a versatile material, its high flammability was a major drawback. A less flammable cellulose derivative, **cellulose acetate** had been discovered in 1865 but it was about 40 years later that an industrial process for its manufacture was found. The Dreyfus brothers made cellulose acetate lacquers which were used to waterproof and stiffen the fabric covering of airplane bodies and wings, especially during World War I. After the war they used lacquers to make Celanese, an acetate rayon yarn. But it was not until the '20s that cellulose acetate molding material

Above: Dr. Leo H. Baekeland.

became available. This new thermoplastic spurred the development of injection molding which soon became the major mass production method for plastics molding. Around the turn of the century another plastics material, less flammable than Celluloid, was discovered in Germany. It was based on **casein**, the protein precipitated from skim milk, and was called Galalith (milk-stone). It also became known as Erinoid, Lactoid, Aladdinite, and Nevalyte – the last presumably because of its nonflammability. Until the '60s this was the main plastics for buttons, buckles, and knitting pins and has been deservedly described as the most beautiful plastic. Many fountain pen and pencil bodies were made from casein as were colorful plastic and bird character creations.

THE DISCOVERY OF SYNTHETIC PLASTICS

The first fully synthetic plastics material was discovered by Belgian-born scientist Leo Baekeland working in the U.S.A. In 1907 he published his first major patent to reveal how two common chemicals, phenol (carbolic acid) and formaldehyde (formalin) could be reacted in the presence of a catalyst to form a treacle-like resin.

He called his material Bakelite and described it as "a material of a thousand uses." This was no exaggeration because he was able to show how molding powders could be produced from the resin by mixing with woodflour and other fillers; how the resin could be cast into open-topped molds for making pipe stems and umbrella handles or mixed with substances such as asbestos and carborundum to make brake linings for vehicles and grinding wheels. Just as importantly, his resins could be used to impregnate papers and fabrics to make tough, electrical-resistant laminates for the emerging radio, telephone, and automobile industries.

As so often happens, a British inventor was working along the same lines. Sir James Swinburne FRS was a great electrical engineer and was searching for new insulating materials for covering wire and cable. He also investigated the chemical reaction but his patent was preceded by Baekeland's by just one day.

Baekeland's work, however, was far more complete and in the late '20s the two men got together to exploit Bakelite in Britain and eventually formed the company Bakelite Limited with a large new factory in Birmingham.

✪ ✪
Above: These butter and margarine holders were produced in polystyrene.

✪ ✪
Above: Dainty and witty statement in Celluloid. A salt-and-pepper shaker in the shape of opera glasses.

Below: This children's music box in the fashion of Mom and Dad's phonograph dates from the '50s. It has the extra bonus feature of little colorful disks in the shape of long-playing records, each of which causes a different tune to be played. Fun for the under-5's.

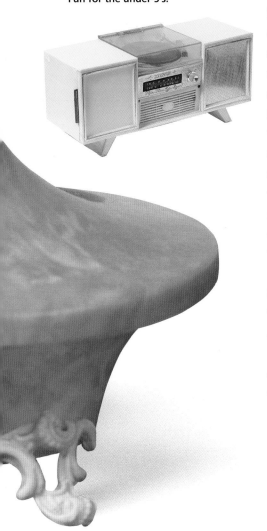

A limiting factor with phenolics was their rather somber tones and small range of colors. Overcoming this problem led researchers into the next milestone in the plastics story: the development in 1926 of light-colored thermosetting resin based on **urea, thiourea** and **formaldehyde.**

This new resin was capable of an almost unlimited range of opaque and translucent colors, and could directly replace Bakelite molding material for most applications where color was required. Molded picnic sets, kitchenware such as eggcups, beakers, and bowls in colors simulating marble or alabaster were given extraordinary trade names such as Beatl, Bandalasta, and Linga-Longa. Later, thiourea was omitted from the urea-formaldehyde mixture which was then only produced in plain and speckled colors. In the '30s, another similar material, **melamine formaldehyde** was made. This had a harder surface and was therefore better suited to decorative laminate surfaces and tableware.

THE PLASTICS REVOLUTION

The world hasn't been the same since the introduction of early plastics such as Bakelite and Celluloid. To match the rigorous demands of a rapid growth in mass consumption as consumerism rampaged during the twentieth century, manufacturers had to find other ways of making their products. It was no good chopping down trees, digging ores from underneath mountains, or catching wild animals and pulling their antlers out to make combs, pens, and telephones. You had to be able to make or synthesize the raw materials and then press, mold, roll, cut, and shape them into objects, whether domestic or industrial, then market, sell, and distribute them to customers at the best price.

The foundation of the great international industry we know today was laid in the years between the two World Wars and just after–sometimes known as "the poly era." It arose as a result of the work of great chemists such as Hermann Staudinger and Wallace Carothers, who conceived and developed the concept of giant molecules. It was their work which led directly to the modern plastics. Most familiar of these to most of us are vinyl, acrylic, polystyrene, polyester, nylon, and polyethylene.

Vinyl, more correctly known as poly(vinyl chloride) or PVC was introduced as a flexible plastics material during the

Above and far left: It is quite beyond comprehension what was going through the mind of the person who thought that such things as seashells and flowers suspended in resin blocks and molded into paperweights would prove to be successful lines of merchandise.

Left: A Celluloid powder bowl. It can be used for face powder or talc.

'30s and developed for the manufacture of unsupported film. Due to a shortage of rubber in World War II, vinyl was employed as a rubber substitute for cable covering. After the war it was stripped from unwanted cables and used to make the notorious plastic raincoat, known in England as the "plastic mac." A technique of producing vinyl plastics from a paste of polymer and plasticizer by gelling on a hot surface originated in Germany. It was developed to produce vinyl-coated materials for upholstery by spreading paste onto a fabric and also used to make flexible "squeaky" toys and dolls, play balls, and other hollow items. A more rigid material for molding LP gramophone records was introduced in the late '40s.

Acrylic was initially developed for the transparent interlayer in laminated "Safety" glass primarily for automobiles, but its application as a plastics was immediately recognized. Shaped aircraft canopies during the '40s were an important product. After the war, its transparency and optical properties were exploited in advertising displays and lamp shades. Many household articles were also manufactured from clear and colored acrylic sheet after being heat-formed into various shapes and twists. Acrylic was available initially only as cast rod and sheet, molding materials being introduced much later.

Although **polystyrene** was discovered in 1839, it was not until almost 100 years later that a commercial method of manufacture was developed. Earlier attempts had been foiled by poor stability, largely because of impurities and a poor understanding of the chemistry involved. Polystyrene is a brittle plastic which has a "tinny" sound when tapped. Increased production for the war effort led to a rapid growth in styrene-based plastics in the immediate postwar years.

Nylon was a direct result of the researches of Wallace Carothers, initially as textile fibers and toothbrush bristles but later as a tough molding material. It was first produced in 1939 and became one of the war's most precious commodities; nylon stockings were among the black market's most valuable lines of merchandise.

Polyethylene (Polythene) was accidentally discovered by ICI in 1933 but its use as a semiflexible plastics was soon developed. As a low-loss dielectric it was vital to the wartime development of radar. Its application as "Tupperware," squeezy bottles, sauce dispensers, and packaging films is well known.

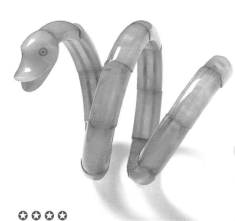

✪ ✪ ✪ ✪
Above: A range of unusual bangles. These bangles with polka-dot inserts are very sought after by collectors.

After the war in Great Britain, J.T. Dickson and J.R. Whinfield of the Calico Printers Association figured out how to combine elements of coal and petroleum into threads thus creating Terylene, the first polyester fiber. In 1950, DuPont purchased the patents, and within just two years they had coined the term "wash and wear." The golden age of man-made fibers had begun. In recent years, polyester has become familiar as one of the materials used for plastic drinks bottles and other food packaging.

Texturized polyester garments were originally a big hit, not only because of their convenience but also because they seemed rakish. Polyester pantsuits, lounge suits, and slacks were made to be fun, in crazy textures and bright, gaudy colors. They were meant to project a sporty, informal image. Knit-look, powder-blue pants flared into grotesquely wide bell-bottoms; ladies' pantsuits of nubby sienna-red polyester knits were available with matching globe-head caps and jumbo pocket books. The world was filled with outrageous clothes that never needed pressing.

But polyester's days were numbered. Discount tables and rails in swank department stores grew overcrowded with polyester apparel, most of it poorly designed and cheaply made. It was the aggressive informality of the designs combined with the fact that pure polyester clothes did not breathe, causing the wearer to sweat profusely, that suddenly caused polyester to appear far from stylish. It became the fabric of the cheap and sweaty, and it still is.

Between 1945 and the '70s, the plastics industry grew at the phenomenal rate of 15 percent per year into a multibillion-dollar industry. In 1982, plastic production surpassed that of steel worldwide, and we formally entered the Plastic Age.

From the very start, plastics have been contradictory substances—mysterious in their elemental origins as organic chemicals and banal in their sheer ubiquity. In both the worlds of industrial design and decorative arts, the role of plastics has evolved from being a mere substitute raw material to that of an essential ingredient responsive both to human needs and fantasies. Plastic as a material now has applications that touch every part of our lives.

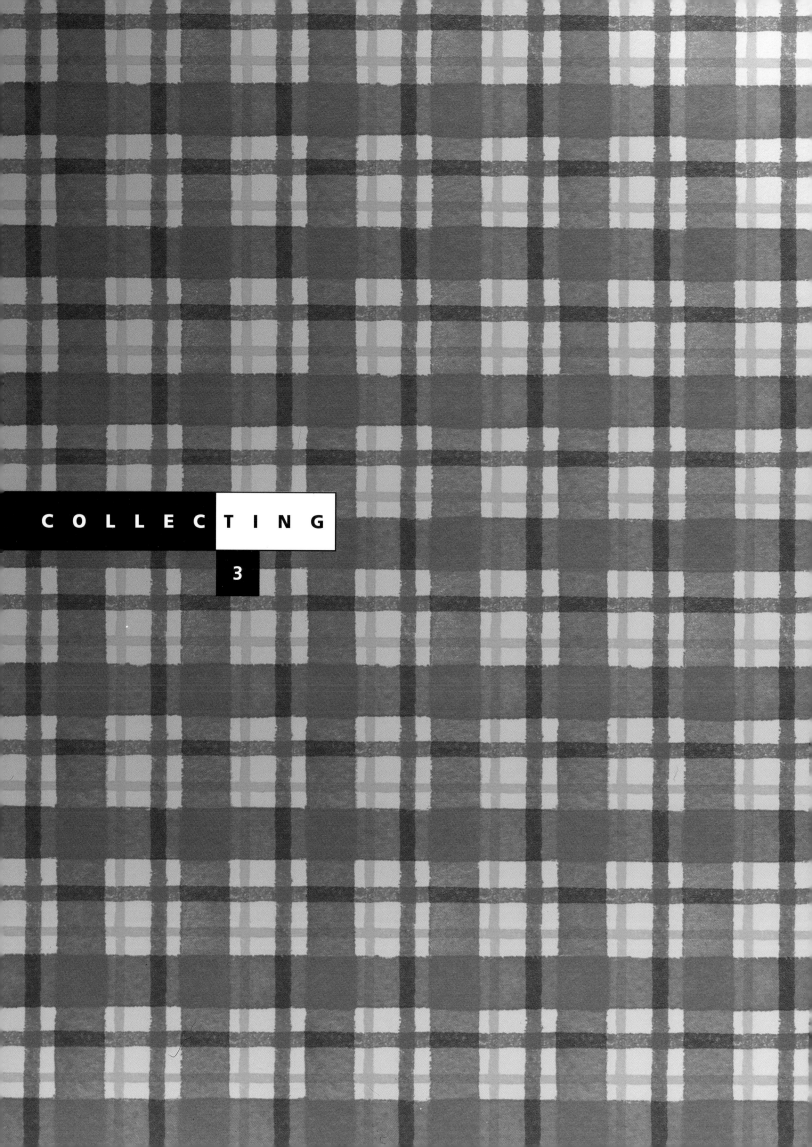

COLLECTING

3

O bsessive collecting, it has been said, is a form of autism. Autistic children are unable to use language meaningfully or to process information from the environment. About half of autistic children are mute, and those with speech often repeat only mechanically what they have heard. The term "autism" refers to their vacant, withdrawn appearance. Other characteristics of autism are a fascination with objects, a ritualistic response to environmental stimuli, and a resistance to any change in the environment.

I only make this point because there is more than a fair element of truth about obsessive collectors and autism. Each and every one of us probably knows an obsessive collector. They seem preprogrammed to rise at the crack of dawn on a weekend morning while the rest of us enjoy an extra hour or so in bed. They follow the same ritual of attending rummage sales or traveling across country to some fair, and appear totally absorbed in their single-minded and endless quest for a purple troll or a Popeye Pez dispenser. If you question any obsessive collectors about their all-consuming hobby, they will almost certainly gaze back at you in stupefied confusion. Collecting IS their lives, their raison d'être, if you will, and who are you to be so impertinent?

But it need not be the same for all of us, we don't all have to devote our entire lives to the fickle quest. Starting a collection can be a personal statement, satisfying a creative urge. Collections can be arranged artistically and give the world a sense of who you are as an individual. There are many benefits from becoming a collector, specifically the thrill of the hunt as you seek new additions for your collection, whatever it may be. Quite often the story of how an object has been obtained can be more interesting than the object itself. The exclusivity of a collection, not many other people collecting what you choose to collect, can be important, as well as meeting and keeping in contact with other collectors. These are just some of the reasons to start, but remember, keep it under control: collecting can be time-consuming and exhausting work.

The best and cheapest places to start collecting plastic kitsch are rummage and garage sales. You can pick up goods cheaply and, if you look hard and long enough and set your standards high, you can get quality artifacts in good condition. By

✪ ✪

Above: Toy violin from the mid '50s with original box adding extra value.

✪ ✪

Left: Various plastics are used in this duck-shaped clothes brush. A few lucky collectors have the plastics egg which nestled in a hollow in the base.

✪ ✪ ✪

Right: This '60s money box is one of the earliest made in polythene.

KEY TO PRICE GUIDE

A price guide appears above every caption throughout this book:

✪	*Do not part with any more than a few dollars*
✪✪	*Has a value of between $10-$20*
✪✪✪	*Worth in the range of $20-$50*
✪✪✪✪	*Highly prized collectible or rare; pay up to $100*
✪✪✪✪✪	*An absolute gem of an artifact that could cost you anything above $100*

starting at the lower end of the price register, you can experiment with collections. Say, for example, you thought that a collection of plastic goldfish is for you, and you're lucky enough to buy a mixed assortment at a good price, but never see anything like it again. Then you quite painlessly abandon the idea and move on to assembling the world's best conglomeration of resin seashell paperweights in puce and turquoise. Collections can be rewarding–accumulate several collections simultaneously–so that it will be unlikely that you will return home empty-handed from any shopping trip.

As your interest in your collections develops, it is inevitable and essential to meet like-minded souls, if only not to feel like a social outcast. More importantly, the potential of swap-meets is of paramount importance to the collector who is becoming quite serious about his hobby. The rapid growth of the Internet has in recent years effectively shrunk the entire world into your home PC. No longer do you just have to eagerly await five and one-half months for the next edition of Nodding Dog Collectors Bi-Annual. Keeping in contact with other collectors is as easy as ABC. Virtually anything collectible has a forum or newsgroup on the Internet and it's amazing how the home-made industry of swapping is thriving in cyberspace.

From here on in, it starts to evolve like any other business. There are mail-order companies specializing in all manner of kitsch plastic merchandise, such as optic-fiber lamps, snow globes and McDonald's giveaways. There are bargain stores and stores big and small in every major town and city specializing in second-hand objects, artifacts, furnishings and fashions from every year, and many of these retail outlets supplement the old with the new by selling new renditions of old classics to those with the discerning eye for the attractively repellent merchandise that is known as kitsch.

Prices range from the fabulously cheap to the offensively expensive. This is always in direct proportion to supply and demand. Remember this is not the conventional retail landscape. There is little competition from an adjacent chain store. If pink fish-shaped basketwork handbags are rare and unusual and they are genuine '50s and in mint condition, you are going to have to pay for it.

CARING FOR YOUR COLLECTION

No one is going to become bankrupt or heartbroken if a minor item in a kitsch collection falls apart through misuse or neglect. On the other hand, most of us have particular favorites which

✪ ✪ ✪
Above: Early '50s ashtray in urea formaldehyde depicting a man worse for wear, clinging onto a lamppost for dear life.

✪
Above: This celebration of the vegetable world is arguably kitsch at its best.

we would wish to maintain in good condition.

Contrary to popular belief, plastics are not indestructible. For example, natural and artificial light will eventually damage all plastics. Excessively high temperatures or humidity can also cause problems with many plastics so seek advice if you have anything of value.

Many objects, when bought, will be dirty or even damaged or deteriorating. It is especially important to examine every object carefully for signs of deterioration because one such object can "infect" a whole collection. The signs to look out for are:

1. Crazing or cracking—particularly in the middle of transparent sections.
2. A "tacky" or oily surface, especially if accompanied with distortion.
3. Discoloration, especially if patchy.
4. Corrosion of metal parts or hinges, etc. which are part of the object.
5. Unusual smell, especially if acidic or vinegary.
6. Excessive brittleness or fragility.

If in doubt about any object, then isolate it from the rest of your collection, as a precaution.

Because surface contamination is often a contributing factor to premature decay it is advisable to clean objects before adding them to the collection. In general, most plastics can be safely washed in warm (not hot) water; indeed, their ability to be easily cleaned has been one of plastics' strong points, from early Celluloid collars and cuffs to modern kitchen surfaces. A few drops of liquid detergent should be added to the water and the use of a soft brush will help remove grime from crevices. Do not allow to soak—rinse in clean water and wipe dry immediately with an absorbent cloth or tissue. Occasionally, however, washing should be avoided altogether. With vulcanite (sometimes called ebonite or hard rubber) that is discolored, washing may make things worse. Be careful, too, with plastics surfaces that have been dyed, especially those which have been cut back to show an underlying color to give a two-tone effect. Porous materials and painted objects also need careful treatment, as do those which contain metals that might corrode. Stubborn dirt can sometimes be removed by a fingernail softened in warm

water. This is softer than most plastics and should not mark them. A wooden cocktail stick is good for removing small spots. Try to avoid using solvent, but cigarette lighter fuel may be used sparingly to remove tar or sticky label residues. Do not use solvent on painted or transfer decoration. Cellulose acetate and plasticized (flexible or semiflexible) vinyl should not be treated with solvent nor should rigid thermoplastics such as polystyrene or acrylic as these may craze. If in doubt, seek expert advice.

If, in spite of trying all the methods above, an object remains soiled, more aggressive cleaning methods may be called for. For example, Bakelite tends to discolor with age and exposure to light and the surface becomes dull. Application of a mildly abrasive wax polish may improve the appearance and offer some protection against light. To restore color any suitable abrasive polish can be used but discoloration will eventually return to the abraded surface. Thermoplastics should not be wax-polished, particularly Celluloid and cellulose acetate which need to "breathe."

Once you have decided that an object is clean and in good order the best way to avoid problems is to follow these simple rules.

1. Keep objects in the dark and avoid exposing them to sunlight, bright daylight or strong artificial light. Use diffused light for displays.

2. Inspect regularly (once a year) and isolate any object showing signs of deterioration so that it cannot cause harm to others.

3. Make sure that storage areas are dry and ventilated— avoid damp or stuffy conditions or cold areas which might encourage dampness.

4. Don't store objects in sealed plastics-bags, packaging, or other sealed enclosures. Do allow some air to circulate and use color-free (preferably acid-free) tissue paper rather than newspaper to wrap objects.

5. Take care when handling decaying plastics. The surface may be acidic—avoid transferring this from your fingers to other objects.

6. Flexible or semiflexible objects should be supported and not stored under any stress or distortion.

7. Flexible plastics, for example vinyl toys, should not be allowed to come into contact with other plastics (especially polystyrene) as irremovable marks may be formed where they touch.

✪✪✪✪✪
Above and top center: A selection of the different styles of Bakelite radios. These are very collectible objects and many may reach to considerable prices.

✪✪✪✪
Center: Australian Radiolette. It is typical of the skyscraper style of radios that were around in the '30s.

✪✪✪✪✪
Far Left: A post-war Bullet, a Catalin which is very sought after in blue.

POP GOES THE PLASTIC

4

Much of what is instantly recognizable as kitsch dates from the most celebrated decade of design excess—the '50s. This decade saw a radical departure from the austerity of the war-dominated designs of the previous 10 years. The whole world appeared to explode into color and a whole variety of new and different shapes and color combinations materialized alongside the development of the new wonder material —plastic. No longer were there only the Bakelite combinations of reds, oranges, and bottle greens, but an entire palette of hues from which to paint the world: pastel pinks, lemons, baby blues, greens, lilacs, turquoises, and oranges.

✪ ✪ ✪

Above: These fun '60s plastic-coated tablecloths with a Mediterranean theme are becoming harder and harder to find. The objects featured on these prints at that time represented the exotic but are commonplace today, such as olive-oil and vinegar bottles, salad bowls, rustic European cheeses, and soups. Somewhere deep within this nostalgia and naiveté lies its real kitsch appeal.

✪ ✪ ✪

Left: Plastic-coated tablecloths of the '60s combining both a pretty powder-puff pink color with the ultimate kitsch pet, the poodle. Particularly sought after by aficionados of the era.

✪ ✪ ✪

Right: If you look closely enough, you'll see that dog again in this beach scene. Another '60s plastic-coated tablecloth.

All these colors quickly became fashionable, and never mind if they didn't particularly go together. The fashions, furnishings, and ornaments from this decade provide the richest store for the kitsch collector, although their abundance is now in decline (and consequently, their prices increasing).

Once unleashed, the shapes of things went wild, too. The symmetry of the past was lost and '50s design gave the illusion of asymmetrical irregularity derived from a wide-ranging and eclectic set of reference sources. Molecular structures, amoebas, kidneys, and even boomerangs appeared as all manner of things from handbags to coffee tables. There were bold new patterns too: chevrons, zebra stripes, leopard spots, polka dots, leaves, travel scenes, and abstract jumbles that were inspired by abstract art.

The mass production of plastic goods was perfected during this heady 10 years, and a comprehensive range of products, cheaper to distribute than ever before, could now be made. This meant that people could acquire

✪ ✪ and ✪ ✪ ✪
Right: Plastic watches are very interesting and practical items to collect. Here are some examples of both the new and old. The range with the clear straps are from the '70s, but the bubble watches are contemporary and produced with you "the kitsch connoisseur" in mind.

✪ ✪
Below: These brightly colored and imaginative resin rings are very cheap to acquire. They are manufactured in a bewildering array of different colors and finishes and with many imaginative offerings inset, from tiny floral bouquets to eyeballs (fake, of course).

✪ ✪

Bubble watches–fun watches especially hand-carved from the finest poly-plastics for the kitsch enthusiast. They combine gaudy colors, jokey graphics, and a knowing nostaligic nod towards the '70s with the digital displays.

✪ ✪ ✪

Above: These fake tortoise-shell '50s frames have never really gone out of fashion. Today there are many imitators, some with very well-known designer names engraved onto the arms. However, the originals as always are the greatest.

✪ ✪ ✪

Right: Sunglasses came in all different shapes and sizes. These glasses were very popular in the '50s.

✪ ✪ ✪

Below: Another type of sunglass here, the plastic frame is transluscent and was very trendy in the '60s.

✪ ✪ ✪

Left: A selection of plastic sunglasses from the past three decades. These examples of eyewear are fairly easy to locate these days and are fun to wear as well as collect. '50s shapes are distinguishable by their winged rims and glittery resin, the '60s are represented by pop-art patterns, and the '70s by their outlandishly oversized hexagonal shapes and the command to LOOK at me! One small footnote: ultra violet protection may not necessarily be guaranteed.

✪ ✪

Below: Lightweight plastic rendition of stainless-steel frames made popular by Elvis Aaron Presley. Not complete without graduated tobacco lenses. Good for collectors of Elvisiana and funny sunglasses; not so impressive these days if you actually wear them.

almost any object they wanted for a fraction of the previous cost if it had been hewn from bone, carved from wood, or cast from clay.

To oil the wheels of the postwar economic boom, the money to acquire these things became easier to come by. A proliferation of money-lending ideas sprung up, especially the brilliant idea of the installment plan. Now,

✪✪ – ✪✪✪

Above: Plastic watches were being developed as exciting, innovative fashion accessories, such as this radio masquerading as a wristwatch. The color combination reflects the style of the decade.

✪

Right: Attention all lady collectors! Complete your kitsch costume jewelry with a range of fruit and flower earrings. Clip-on attachments will give you some indication of their low price, so will the complete lack of any gemstones. However, these are very wearable and durable.

✪ ✪ ✪ ✪ ✪

Left: Years ahead of his time in the '70s, British inventor and entrepreneur Sir Clive Sinclair pioneered the development of personal computers and a range of stylish electronic goods. There is quite an esoteric bunch of people collecting this sort of equipment. Digital watches.

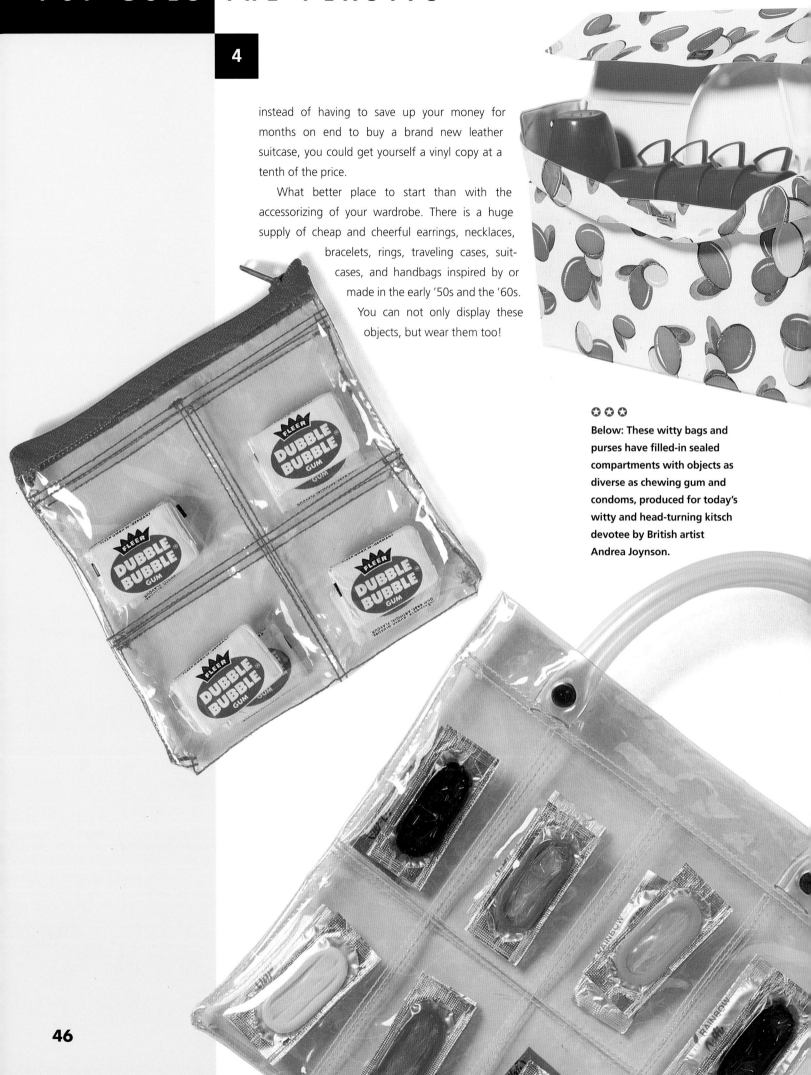

instead of having to save up your money for months on end to buy a brand new leather suitcase, you could get yourself a vinyl copy at a tenth of the price.

What better place to start than with the accessorizing of your wardrobe. There is a huge supply of cheap and cheerful earrings, necklaces, bracelets, rings, traveling cases, suitcases, and handbags inspired by or made in the early '50s and the '60s. You can not only display these objects, but wear them too!

✪ ✪ ✪
Below: These witty bags and purses have filled-in sealed compartments with objects as diverse as chewing gum and condoms, produced for today's witty and head-turning kitsch devotee by British artist Andrea Joynson.

✪ ✪

Left: Picnics were very popular in the old days. So cute '60s picnic sets having a washable vinyl cover with jolly color motifs such as this, were very popular at the time.

✪ ✪

Right: These vinyl purses are new but have that '60s authenticity, without holes from being overused.

✪ ✪

Right: Fun fur is a big favorite with all kitsch enthusiasts. Perhaps it's something to do with the fact that fun fur is just so unlike the real thing. These purses in fake leopard skin (and what looks like domestic cat fur) are being manufactured today.

47

✪ ✪ ✪

Right: This classic airline overnight bag is from Britain's now defunct BOAC airline that once dominated European skies in the '60s. Hence its rarity and unusualness and price.

✪ ✪

Center: Most luggage today is so dull and homogenous, it's a wonder we can tell which bag belongs to us at baggage reclaim areas. Hark back to a more naive time, when international travel was less like an everyday experience. These '60s handbags, overnight cases, and duffel bags in classic print patterns from those eras, depict the excitement of the time at the notion of traveling and visiting foreign countries with their unusual smells and customs.

✪ ✪

Left: This shiny vinyl overnight bag probably dates from the early '60s/late '50s and is a rarity.

✪ ✪ ✪

Right: This patent vinyl overnight case from the '70s depicting a ballerina would have been the height of sophistication at the time.

✿ ✿

Above: Fun fur or fur prints add that extra touch of glamor to any weekend away.

✿ ✿

Center: Now you see it, now you don't. But you'll be able to see what you keep inside. A very contemporary clear vinyl duffel bag.

✿ ✿

Below: These plastic-covered clutch bags that look like rolled up magazines date from the '70s. They are very rare and the ultimate accessory for the '70s fashion freak.

THE GREAT PRETENDER

5

5

Plastic, of course, originally came into being because it could emulate real materials such as ivory and tortoise shell. Nowadays, it seems that through advanced "stealth" technologies, there is no material that plastic cannot pass itself off as. The effects of some of these impersonations are best viewed at a distance. For a few of the examples that you will see in this chapter, it is advisable that you take 20 or so backward steps and peer through squinting eyes to believe that they could be passed off as the real things.

However, some of the advantages that plastic has over natural materials, other than the cost of manufacture, are that in many cases it can last longer, can be more durable than the real thing, is washable, and looks quite (and I emphasize that word) authentic. However, as we all know, plastic can give rise to some rather beautiful examples of kitschy plastic merchandise.

✪ ✪ ✪

Above and below: "Ivory" and "brass" "old-fashioned" telephones in the '40s style which were more than likely manufactured in the '60s. Very popular with nouveau riche the world over.

✪ ✪

Right: Early '70s "glass" lampshades were popular not only because of their cost but also because their weight did not threaten to bring down the entire ceiling.

✪ ✪

Above: Why waste money on brittle and expensive-to-replace handcrafted china ornaments when you can have virtually indestructible ones instead. These "Capodimonte" deer figures were very popular in the late '60s and early '70s.

✪ ✪

Below: "Silver" fish-shaped soap holder dating from the '70s. The big advantage that this object has is that it will not require constant polishing because of the tarnishing effect water has. Goes beautifully with the telephones.

✪ ✪

Below: "Apples" combined with a peach-colored "glass" fruit bowl, both from the late '60s. Available at rummage sales everywhere.

Fur was once only worn by the wealthy and stylish. Although there are probably still a few octogenarian princesses and faded glamor queens who cling onto their minks and leopards, you are most likely to see real fur adorning the rapidly expanding figures of the new vulgarians, or nouveau riche. Wearing the pelts of dead animals is one of the crassest and most politically incorrect manners in which to demonstrate the level of your bank balance.

If you don't have the cash but want to appear flash, there is always fake fur. To much of the civilized world, bad taste is a much lesser crime than the slaughter of defenseless animals in the name of fashion. Fake furs are known in the textile industries as "high-pile fabrics" and can be produced in any variation of pigment (e.g. fluorescent color or polka-dot overprints), length, and nap.

It was the highly coveted pelts of a large African and South Asian carnivorous quadruped that inspired and galvanized the fake-fur industry. Genuine leopard skin was always very expensive, but because the simple black-on-gold spot design was easy to emulate, fake-fur makers made the wild animal look available to everyone. Leopard skin became one of the primary patterns of the '50s (along with the boomerangs, amoebas, and kidneys) because of advances in the manufacturing processes of acrylic fibers. Rayon, Dacron, and modacrylic fabrics could simply be printed with a cream tan background and tawny spots to create the look, if not the feel (or thankfully, odor) of a feral cat.

By the '60s, leopard skin became ensconced as a symbol of trashy, ersatz glamor and became simulated on everything from underwear to linoleum. One of the premier kitsch icons, Elvis Presley, created one of the most impressive displays of fake-fur interior design with his Jungle Room at his Gracelands mansion. It was filled from floor to ceiling with zebra-skin furniture and rabbit-fur pillows all set off beautifully with furry green carpeting that ran all the way up the walls and onto the ceiling. The "King of Rock and Roll" liked fake fur so much, he even had his bathroom scales covered in the stuff.

✪ ✪ ✪
Above: This "bone china" plate set from the Festival of Britain of 1951 is almost a genuine antique, commemorating the national celebration that marked the end of the war years and embracing the technologies of the future.

✪ ✪ ✪
Center top: A "silver" lantern-shaped salt and pepper set. It was very popular.

✪ ✪ ✪
Center: A "silver" strawberry cruet set, which is most unusual, not to say bizarre. This probably dates back to the '70s.

✪ ✪

Above: "Fur" cushions made from "tigers," "zebras," "cows," and "leopards." A versatile and contemporary nod toward the kitsch aesthetic that can be successfully mixed with more tasteful furnishings.

✪ ✪

Left: This starfish looks quite real but is in fact, plastic and looks great in any bathroom. Available in stores everywhere.

✪ ✪

Left: Plastic flowers became popular because they weren't so demanding as real flowers. They didn't die and never needed watering. However, they looked incredibly unconvincing. Perhaps it was because, even at a fair distance, they look cold and hard to the touch, unlike the soft, fragrant feel of real petals. (Only the most recent synthetic silk flowers are so realistic that even after you have touched and sniffed them, you can never be totally sure.) They are wonderfully kitsch items, especially when arranged so thoughtlessly as these examples from the '60s and '70s which here are combined with fake plastic (what else?) basketweave. They naturally fade to the most revolting hues, adding an extra special tawdry kitsch value.

✪ ✪

Right: You might be fooled, though, by these plastic indoor pot plants, but don't water them; it'll only run all over your floor and ruin your carpet.

✪ ✪

Below: A contemporary and witty statement on the classic plastic floral theme rendition—an inflatable tulip.

✪ ✪

Above: Plastic fruit has gained the status of an instantly recognizable kitsch classic. Vividly unnatural colors; hard, shiny-looking skins; and unmistakably fake. Cheap decorative objects for those people who in the '50s and '60s lived off TV dinners and canned food. These were folk who couldn't be bothered to grapple with peeling or, come to think of it, chewing their food. These vines and grapes, however, are cheap, fairly realistic and look great if you buy enough of them to create a lavish display, as shown here.

HOUSE SPACE

6

ome is where the heart is and in every dream home there's a heartache. There's something for everyone from any decade at garage sales and flea markets to brighten up any lounge, kitchen, bathroom, or garden. It can come as cheaply or as expensively vintage or original as you like. You can mix the decades together or just focus on a '50s lounge, with chairs, sofas, tables, and lighting still in good condition. In many cases, you can reupholster or recover in original or extremely good reproduction materials which are being produced again to satisfy the retro-hip markets.

There is a great range of plastic kitsch available from all the decades for the kitchen, plenty enough to create a "kitschen" if you so desired. Plenty of novelty ideas using food and (edible) animals within the design seem particularly daft now.

✪ ✪ ✪ ✪

Above: A multistory, multicolored display stand. This object is classic '50s kitsch combining those unique kidney shapes with pastelized primary colors. These are quite hard to come by in this condition simply because of their age and can fetch quite a fair price at specialist dealers, so look out for house clearances and in thrift stores. This was originally used mostly for showing off ornaments and plants or even offering light snacks at cocktail parties, but to the collector it would make a stylish display for a collection of '50s kitsch objects.

✪

Below: Some ultra-cheap mantelpiece furniture. This collection of sickly sweet sentimentalia from every decade was all picked up for just a few pennies a piece; a happy resin sunflower, a sleepy and doe-eyed pooch, lovable old grandpa, some lovebird rodents, and lovers on a park bench.

✪ ✪ ✪ ✪

Center: Backscratchers used to be very popular and are becoming increasingly collectible. They came in a huge range of different materials and styles.

✪

Far right: Here's an ultra-cheap and handy way of brightening up even the dullest doorway. Modern, but beautiful, the beads look like boiled candy close up. Available at any cheap hardware store.

62

"Pass the salt" is something that ordinary people say to each other every day when they desire to season their food, but if there's a cruet set collector at your table, he or she will probably be more interested in the container than the contents or even the food. If your set is a missing link in a collection, you may well receive an offer on the spot.

The appreciation of kitsch doesn't necessarily need to be restricted to the great indoors. Have you ever considered the notion of garden kitsch? There's an entire range of creatures and objects produced to brighten up yards and gardens. They include happy green frogs reclining on toadstools, dwarfs pushing wheelbarrows, and gnomes snoozing under mushrooms. There are many irrefutably tasteless effigies, such as lazy Mexicans with bare feet, cherubs with fountain-spouting penises, and comical drunks leaning on lampposts.

Most lawn ornaments are sold from sprawling nurseries by highways—acres teeming with fauna begging to beautify your life. Find a lawn-ornament store, and you have found a gateway to the world of bad taste. Their popularity sprung up after World War II hand in hand with suburbs and as mobile-home parks blanketed the land. The lawn-ornament industry stood ready and well equipped to beautify them. Plastic and plaster statues were available in abundance either bare or, best of all, garishly prepainted in dozens of shapes, including the all-time favorite, the pink flamingo. Hot-pink flamingos joined the grass menagerie about the time that Florida was popularized as a vacation destination and retirement spot in the '50s. The exotic-looking bird became an emblem not only of Florida but of the era of stylistic excess.

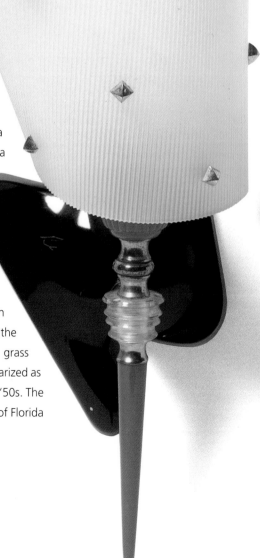

✪ ✪ ✪ ✪
Right: Classic '70s illuminated plastic kitsch to rival the glass, water, and wax lava lamp—the optic-fiber tree. While the evening away watching the strands change color slowly. Reproductions have been surfacing recently, but prices remain consistent whether new or old.

✪ ✪ ✪
Below: '60s "fancy" wall-mounted light.

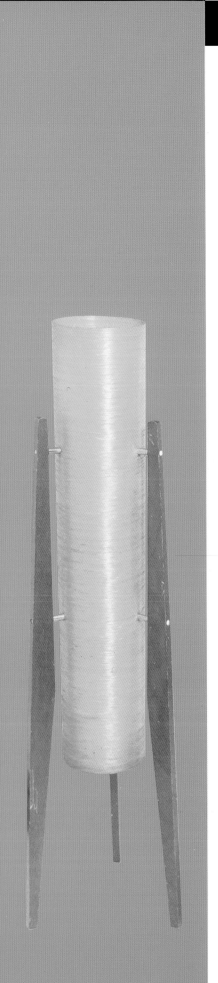

✿ ✿ ✿

Top left: This red plastic piano
is a very handy cigarette
dispenser.

✿ ✿ ✿

Top right: These were very
popular as drinking cups
disguised as other objects. This
one is an ice-cream cone.

✿ ✿ ✿

Center: An money box cleverly
disguised as grandpa sitting in
a rocking chair.

✿ ✿ ✿

Bottom: A very large fountain
pen. It doesn't actually write
but unscrew the bottom and it
is a handy pencil and pen case.

✿ ✿ ✿

Left: These spun-fiber '60s floor
lamps are very attractive and
give off a soothing reddish-
orangy glow.

✪✪

Below: In the mid-'60s when smoking didn't pose the health threat, ashtrays were designed to feature as prominent pieces of furniture in the lounge. If you still partake in smoking for relaxation, this pleasing orange plastic and chrome standing ashtray is always willing to catch your falling ash.

✪✪✪✪

Right: What better way to unwind after a stressful day at the office than to curl up and swivel about in a black vinyl easy chair? These '60s/'70s seats were available in a range of plastic-based upholstery, but the master of them all has to be PVC, such as this example.

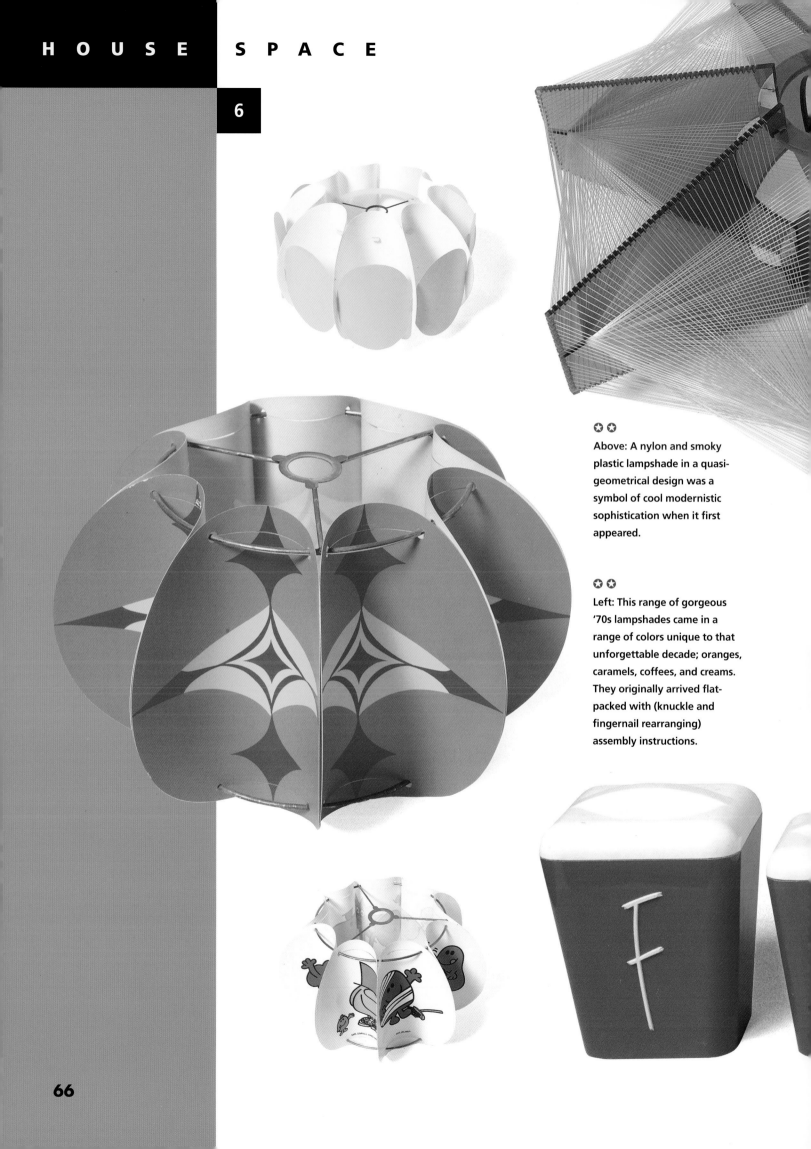

✪ ✪

Above: A nylon and smoky plastic lampshade in a quasi-geometrical design was a symbol of cool modernistic sophistication when it first appeared.

✪ ✪

Left: This range of gorgeous '70s lampshades came in a range of colors unique to that unforgettable decade; oranges, caramels, coffees, and creams. They originally arrived flat-packed with (knuckle and fingernail rearranging) assembly instructions.

✪ ✪

Above: Multi-primary-colored spice racks from the '60s, such as these, are usually inexpensive but quite hard to find in a good and usable condition.

✪

Center: These cute little primary-colored egg cups are quite common even though they originate from the '60s, and should not be too expensive. Always look for the box, if possible, as this will add value and probably ensure that the egg cups will be in good condition.

✪ ✪

Below: This kitchen companion set is a classic early '60s example. "F" is presumably for flour, "S" for sugar, "R" for rice, "T" for tea, "C" for coffee. Any ideas what the smallest canister labeled "A" was supposed to hold?

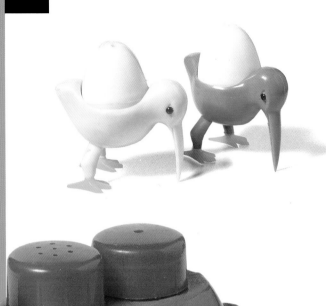

Cruet sets come in all shapes and sizes and are a genre in themselves; indeed there are several books available on the subject and many collectors and societies dedicated to the cause. Because of their abundance, it is a good area to focus on if you start a collection of "kitschen" objects.

✪ ✪

Top: These funny and cute little chicklet egg cups date from the early '60s. If found with original packaging, snap them up immediately.

✪ ✪

Center: This brittle camel has a salt-and-pepper pot instead of two humps full of food.

✪ ✪

Left: These two sad looking hounds are produced in a heavy vinyl material which is doing a poor imitation of china.

✪ ✪

Bottom: The vegetable kitsch icon, the tomato, doubles as a salt-and-pepper set as well as an apt squeezable polythene container for ketchup and salad dressing.

✪

Below: These humorous '50s artichokes are a good example of well-maintained vintage salt-and-pepper sets, but are still relatively cheap. Made in urea-formaldehyde.

✪

Top: A cute Scottish cruet set, lift off their hats to reveal salt-and-pepper shakers.

✪

Right: Two cutesy cruet creatures. Are they bunnies or cats? Perhaps a cross between the two.

70

✪
Right: These '60s polka-dot pastel-colored drinking glasses are still abundantly available today.

✪ ✪ ✪
Below: "Bandalasta Ware" used to be very popular in the '60s. Ideal for picnics and came in a range of colors.

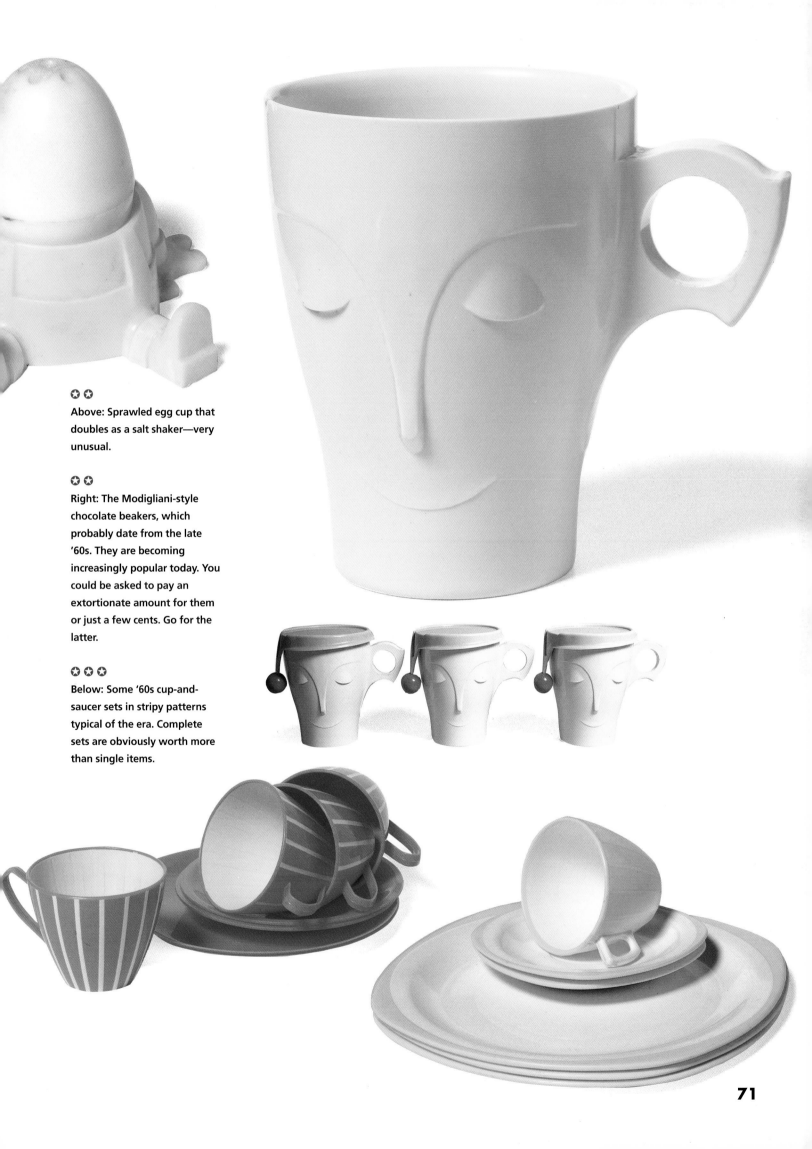

✪ ✪

Above: Sprawled egg cup that doubles as a salt shaker—very unusual.

✪ ✪

Right: The Modigliani-style chocolate beakers, which probably date from the late '60s. They are becoming increasingly popular today. You could be asked to pay an extortionate amount for them or just a few cents. Go for the latter.

✪ ✪ ✪

Below: Some '60s cup-and-saucer sets in stripy patterns typical of the era. Complete sets are obviously worth more than single items.

SALAD SET

SALAD FORK & SPOON
WITH
SALT & PEPPER SHAKERS

SALAD SET

MADE IN HONG KONG

 ✪ ✪ ✪
Above: A '60s salad-server set with box and "wood" and "stainless steel" cruet set.

✪ ✪
Above right: Weird '70s egg-timer set in triangular pink resin block.

✪ ✪
Center: These are fun animal-shaped napkin rings. Useful for any occassion.

✪ ✪
Left: Snack trays available in geometric patterns or illness colors.

✪ ✪
Top: Mustard dispenser with original box.

✪ ✪ ✪
Above: Vinyl-covered floral dining chair from the '60s.

✪ ✪ ✪
Center: Very attractive '60s custard yellow and black salad-server set.

✪✪

Above: These modern jolly and mischievous plastic gnomes have several advantages over their plaster-cast peers: they're virtually unbreakable, their paint won't chip or fade as badly, and they clearly enjoy their lives and therefore will probably live longer. Available as vintage or new models. Both fetch the same price, as years of battling with the elements will have devalued nearly all the oldies.

✪✪

Center: These are wonderful containers to grow garden bulbs. They would look great on any windowsill or patio. Bulb growers come in all different shapes and sizes and are great to collect.

✪ ✪

Right: This was in its day a fairly standard desk phone, but this beautiful '70s hue of orange makes it stand out as a collector's piece today.

✪ ✪

Left: These brightly colored windmills have, in fact, a practical application: they are mole-scarers. Simply pierce the ground with their metal stems and let nature do the rest—the vibrations down the stem by the whirring windmills will keep all troublesome moles away. They can be used in combination with a plastic floral display indoors or even on their own as rather strange conversation pieces.

✪ ✪

Center: Wake up to the smooth and satisfying (and muffled) sounds of AM radio with this vinyl-cased digital alarm-clock radio manufactured by the Japanese company, Sanyo. Probably early '70s. Doubles as dice too. Versatile.

✪ ✪

Left: Manufactured by a leading telephone company, Ericsson, these classic upright telephones were revolutionary when they first appeared in the late '60s/early '70s.

76

Orange plastic sets are quintessential symbols of '70s design. Geometric symmetry was a pleasing aesthetic quality at the time as were curves, waves, and chevrons. And, of course, that color, due to the effect of passing time, now, not quite the bright and sparkling citrussy shade but just plain old orange. They are reasonably easy to detect but tend to be found in bits and pieces rather than as complete sets. If an original box is also present, this will as always cause the second-hand value to soar.

✪ ✪ ✪

Above This layered plastic party set is both versatile and quite unusual, an extra-added value with original box.

✪ ✪

Center: Combs can come in a variety of shapes and sizes. Here, this comb is in the shape of a foot. Very popular in the '80s.

✪ ✪ ✪

Left: This wipe-clean interlocking nest of occasional tables dates from the '70s. Unusual; also popular in white, cream, and coffee.

✪ ✪

Below: These orange egg cups come with their own built-in tidy tray.

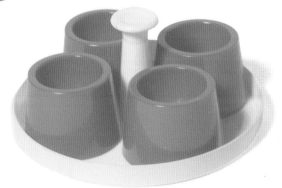

✪ ✪

Above: Everything you'd ever need for those coffee mornings: cups, coasters, and tray. Only the glassware is not plastic.

✪ ✪

Right: Delightful and daring candlestick holder in glass-look clear polystyrene, in orange naturally.

✪ ✪ ✪

Left: This geometric, interlocking drinking glass, jug, and stirrer arrangement suffers from some of the problems that were typical of '70s design: unnecessary features such as interlocking parts, inadequate insulation from hot liquids, and awkward shapes to handle. However, it would have appeared to be the height of fashion then.

✪ ✪

Right: Quintessential '70s design: fake glass lamp stand and a shade in the unique fall colors of that decade. Also has the added bonus of a wipe-clean plastic coating.

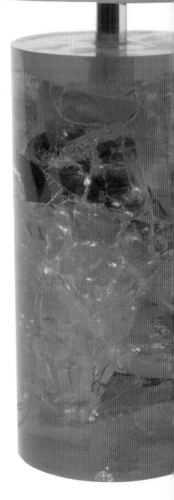

✪ ✪ ✪

Top: A '50s Morphy Richards hairdryer. More of a design-museum piece, a wonderfully significant representation of the influences that automotive and aerodynamic designs have on everyday household goods.

✪

Right: A very handy soap holder in the shape of a bathtub. No bathroom should be without one.

✪ ✪

Right: Imitation mother-of-pearl toiletry accompaniments. They fade into really vile tobacco-yellow colors. Truly revolting.

✪ ✪

Left: This "frosted glass" night lamp will soothe you into gentle slumbers until you look at it and realize it's just a plastic rendition of frosted glass.

✪

Bottom: A handy calendar and barometer in polystyrene. Popular TV-top accompaniment from as far back in history as the '60s.

✪ ✪ ✪ ✪ ✪

Opposite page: This is one of those instantly recognizable kitsch classics. A '50s boat bar carved from "beech" and "rosewood."

Living-room cocktail parties were a popular social feature in suburbia in the '50s and '60s. Neighbors cooked barbecues, played cards, and generally had a good look at how one another lived. It was an era in which the notion of social status took hold and status symbols became all powerful. The sense of fun of the time is present in the designs of these social tools.

✪✪✪✪✪

Above: Barrel bar made by plastic coopers in "teak" and "brass."

✪✪✪

Above: A '60s sparkly vinyl-covered bar stool featuring fur sparkly patterns with astronomical motifs.

✪✪

Above: Liven up your vodka martinis with these vividly-hued plastic "glasses" from the '60s.

✪✪

Left: Collage-pattern drinks-related products: ice buckets, drinking glasses, and cocktail shakers. Very '60s pop art and quite unusual.

Above: Golfers' ice buckets for that extra ice at the 19th hole.

Above: Smokers' complete set from the '60s: ashtray, cigarette dispenser, and lighter, all on a soccer-ball theme.

ZULU-LULU WILL MAKE YOUR GUESTS "BUST" OUT LAUGHING!

LOOK what a few years do to LULU!

Nifty at 15 · Spiffy at 20 · Sizzling at 25 · Perky at 30 · Declining at 35 · Droopy at 40

The Newest SWIZZLE STICK SENSATION...

Don't pity Lulu — you're not getting any younger yourself... laugh with your guests when they find these hilarious swizzle sticks in their drinks. ZULU-LULU will be the most popular girl at your party!

MADE IN HONG KONG

Style sexy Drinking Straw

Above: Hours of krazy cocktail fun to be had with this variety of propositions. Far more fun than others, particularly the Kama Sutra set (see above).

✪ ✪
Left: A "fruits of the forest" range of fruit-shaped ice buckets from the '60s and '70s. Very popular with collectors, quite easy to locate, better value if matching drinking glasses are available (below).

✪ ✪ ✪

Left: You need to look closely at this cigarette lighter to see tiny goldfish set in the resin.

✪ ✪

Right: This handy string dispenser now appears to be deeply racist.

✪ ✪ ✪

Below: A '50s snack tray featuring a playing-card theme. Ideal for nibbles for those cocktails.

✪ ✪

Bottom: Finally, a brilliant idea for the frequent traveler: an inflatable coathanger.

BRAND **NAME BEAUTIES**

7

By the '60s, virtually every home had a TV set and, therefore, TV commercials with characters hawking everything from breakfast cereal to bubble baths. The typical advertising representative on TV before this era was considered to be a well-groomed smiling guy carrying an attaché case. In the '60s, he underwent a pretty major personality transformation, and there appeared a new breed of rep.

These new reps no longer had flesh or blood coursing through their veins, they became two-dimensional cartoon-like characters. Talking dogs and cats, giant fish, men made of dough, and many other anthropomorphic shapes seemed to serve better.

Recognizing that these new TV admen had become as popular as the products they hyped, companies began to produce three-dimensional replicas, mostly in vinyl and other plastics. Today these spokesbeings have become highly collectible.

✪ ✪ ✪

There are several books now available solely devoted to McDonald's collectibles. Here's a very rare and very old McDonald's piece of merchandise: a drive-in snack tray. Photo courtesy Hake's Americana, York, PA.

**Ronald McDonald has been the official clown prince for McDonald's since 1963. He is probably the world's most instantly recognizable advertising character. As such he reigns as the identifying symbol over the vast global empire of fast-food restaurants and the even greater assortment of McDonald's giveaways of McDonald's happy toys.
With 15,000 McDonald's locations in over 70 countries, collectors are everywhere. To help them find each other, The McDonald's Collector's Club was established in 1990. It initially had just 18 members, but today it has expanded to over 1,200 fans in the USA, Canada, Germany, the UK, Holland, France, Ireland, Australia, New Zealand, and Singapore.**

Disney character © Disney Enterprises, Inc.
Used by permission from Disney Enterprises, Inc.

Here is just a small selection of what's available and an example of how attractive such a collection of freebie giveaways can look with just a modicum of care. You don't have to pay for any McDonald's toys as they are given away free as promotional goodies mostly with kids' meals. However, they soon appreciate in value and, although very common at garage sales and church fairs, certain rarities can fetch a pretty cent at specialist conventions.

Here are some simple buying and storing tips that should apply to any plastic collector:

- Try not to buy anything with missing pieces, through either chipping or breaking.
- If the object had a function (e.g., a pencil sharpener or an ashtray), it should still be able to fulfill that function.
- If you think that your collection will seriously increase in value and your main motive is one of investment, you should consider wrapping your beloved items in paper and storing them in the dark. But I think that there's little point in owning them if you can't proudly display and enjoy these objects. They will stay in good condition if kept out of direct sunlight and away from sources of heat, and from children and water.

Above and right: These chimpanzees have been hugely popular in Great Britain for nearly 30 years. Real chimpanzees dressed as people voiced over by humans have starred in an endless series of British comical TV ads for PG Tips, a favorite brand of the nation's favorite drink, tea.

Left: The Esso oil man (here complete with oil) was a favorite in the '60s.

✪ ✪ ✪

Below: The world-famous Michelin man who endorsed the company's range of car and truck tires is another instantly recognizable and lovable corporate symbol like Ronald. Many are from car-part stores and garages and, because of their size, can be adapted for other uses, such as lamps.

✪ ✪ ✪

Above Left: Poppin' Fresh Popcorn cruet set. Also known as the Pillsbury Doughboy and his girlfriend, Poppie. These plump, happy figures derive their names from both the baking dough and the sound it makes when a tube of Pillsbury refrigerated dinner rolls are opened. Doughboy first appeared in 1965. Because of his immediate popularity as a spokesperson, he took on a three-dimensional form as a doll, a clock, a cookie jar, a radio, and even a telephone. Pillsbury even produced other family members; look out for Popper, Bun-Bun, Grandmommer, Grandpopper, and Uncle Rollie. Doughboy also had two pets, Cupcake and Dough-Bear. Photo courtesy Hake's Americana, York, PA.

✪ ✪

Above and Below: A variety of California Raisin's promotional cuties including inflatables, key rings, and tiny wind-up walking toys. Photos courtesy Hake's Americana, York, PA.

Pez dispensers are a familiar mainstay of the checkout aisles of supermarkets across America. They offer great temptation to the impulse buyer, tucked away between the candy bars and key-chain flashlights. The recent rise in popularity of these colorful plastic gizmos is reflected by the appearances or mentions in recent hit TV series such as *Married With Children, Cheers, Thirtysomething*, and *Murphy Brown*. One episode of *Seinfeld* revolved around the Tweety dispenser. And when prestigious Christie's auction house included several lots of Pez dispensers in a recent auction, it became clear that the candy container was a main contender in the market.

Pez has been around a lot longer than most people imagine. The candy was invented in 1927 by Eduard Haas III, an Austrian food merchant. Haas was a creative businessman with detailed knowledge of the properties of peppermint oil. He initially created the small rectangular candy as a breath mint. Advertised as an alternative to smoking, it became very popular very quickly, fashionable almost. Haas named his brick-shaped candies Pez (because it was pronounceable the whole world over) by using the first, middle, and last letter in the German word for peppermint: PfeffErminZ.

The incorporation of the characters' heads onto the dispensers came about with a shift in marketing in the early '50s toward children. Along with the inclusion of fruit flavors, the collectible heads featured holiday and popular, licensed characters. Among this first generation of characters were One-Piece Witch, Easter Bunnies, Santa, Popeye, Mickey Mouse, and Donald Duck.

There have been over 300 different heads manufactured and available to the public since then with at least 75 different dispensers currently available. The variety of heads available may account for the popularity of Pez, with almost 50 different Disney heads, Looney Tunes, and numerous superheroes and villains. There are monsters, astronauts, historical figures, animal heads, and trucks. An impressive number of licensed characters have made it to the top of the Pez dispenser including Garfield, the Muppets, Peanuts characters, Smurfs, and the Flintstones.

Disney character © Disney Enterprises, Inc.
Used by permission from Disney Enterprises, Inc.

✪ ✪ ✪
Below: Khan's "Beefy Frank" mustard bottle celebrating the famous brand of hot dog. Photo courtesy Hake's Americana, York, PA.

✪ ✪ ✪

Left: '60s Alka-Seltzer bank. Speedy is the personification of digestive relief promised by the age-old aspirin and bicarbonate of soda tablets manufactured by Miles Laboratories. Red-head Speedy was born in 1951 and holds an Alka-Seltzer tablet to his chest and wears another as a hat. Speedy became a household name in cleverly animated TV commercials from 1951 until he was pensioned off to retirement in 1964. Photo courtesy Hake's Americana, York, PA.

✪ ✪

Below and Right: Pez dispensers—Captain Hook, Mr Ugly, Sheriff, Gorilla, and Jiminy Cricket.

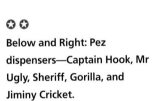

PLASTIC PLAYMATES

8

94

Once upon a time, the world of toys was a much simpler one. Children played with wooden abacuses and rocking horses, rag dolls, marbles, and metal trains and cars. Then along came plastic and bang! In the '50s the toy industry exploded into life. The powerful combination of television advertising and children's serials made the invention of new toys into an incredibly big business. All the major toy companies grew into the monsters they are today—Hasbro, Mattel, Lego, and Fisher Price to name but a few. Superpower toys such as Barbie, Frisbee, and Etch-a-sketch all emerged in the '50s and '60s.

The toys that were popular from this era are still extremely popular today. For the kitsch plastic collector, the originals will hold the most fascination but so will the prices of vintage toys. There is an enthusiastic and serious market simply devoted to vintage toys and a lot of collectors and fairs the world over, so if you're operating on a budget, you may have to specialize in collecting toys that aren't very popular with toy collectors. However, when you are trawling garage and rummage sales, keep your eyes open for a bargain because you could sell a vintage toy to a serious collector and make a profit!

✪ ✪ ✪

Below: An unusual harmonica in the shape of a bright yellow banana. Available from the Canary Islands.

✪ ✪ ✪

Left: A relatively old model of a Cardinals player. Now quite rare and very collectible. Photo courtesy Hake's Americana, York, PA.

✪ ✪ ✪

Right: Coca Cola have produced a whole range of items including this highly prized delivery man doll. Quite rare and very collectible.

✪ ✪ ✪

Below: Would you allow your children to play with any of these? This vaguely grotesque collection cost virtually nothing from a rummage sale. A mixture of the evil, the strange, and innocence epitomises most of these very cheap creatures. From left to right: a purple push-puppet cat (if you push the base in, the cat collapses in a boneless heap), a happy purple cow squeaky doll, a Little Red Riding Hood rocker, a pink bunny wobbly toy and a false-looking rabbit.

✪ – ✪✪✪✪✪

Mr. & Mrs. Potato Head pioneered the road to success for the plastic toy industry in the '50s. In 1952, the Hassenfeld Bros. (later Hasbro) introduced this unassuming toy around the world. Significantly, it was one of the first toys ever advertised on the all-new powerful medium of television. A chintzy black-and-white TV advert showed a band of kids sticking plastic ears, noses, mouths, and eyes into real-life potatoes. Unlike the slick-looking kids in TV commercials today, the children in the Mr. Potato Head ads got everything wrong. Mr. Potato Head's face was askew, and they forgot to put on his nose. All they did was have fun and overnight he became a sensation.

Most collectors find it hard to leave Mr. Potato Head sitting in the box. He's so kitschy and fun, they find themselves rearranging eyebrows, ears, and his trademark pipe. Since the mid-'60s, the kits have come with a plastic spud; this helps date the kits. Original sets had all the body parts: a hard plastic body and shoes, as well as all the face furniture. The great thing about Mr. Potato Head is that there are so many different types. For under $20 you can sometimes get loose potato head parts without their original box, but prices go up very sharply for complete sets, particularly those from the '50s which can go for as much as $200. At the very top of the price range, the primo spud man is Mr. Potato Head on the Moon, price by negotiation. Cookie the Cucumber photo (right) courtesy Hake's Americana, York, PA.

✪ ✪

Above: Early '70s children's do-it-yourself plastic craft kit. Enabled schoolchildren to produce unique but truly ugly objects themselves.

✪ ✪

Right: New Tintin dolls are expensive items already. You may be best advised to look for tattier older models in rummage sales.

✪ each (or ✪ ✪ ✪ for the house)

Left: Smurf toys were hugely popular in the early '80s. These strange little blue-colored elves were corralled by an elderly and avuncular rabbi called Father Abraham. They have made a recent comeback bid by making a record of cover versions of popular chart hits.

101

✪ ✪ ✪

Left: This '60s doll of air stewardess is from the little Irish airline, Ryanair. It is a rarity because of the size of the airline.

✪ ✪ ✪

Right: A selection of kitsch toys from the '80s. Dancing flowers, at this time, were very popular. They furnished many households. Just switch a button and the flower will dance wildly. An interesting conversation topic at any dinner party.

Prices range from
✪✪ to ✪✪✪

Right: There was a time when Trolls were simply quiet mischievous characters from Scandinavian folklore. Living under bridges or in caves, they were known for being dirty, ugly, and consummate pranksters. Trolls stole food, created toothaches in children who did not brush their teeth, and matted the hair of children who forgot to brush it. Legend has it that if you caught one of the little creatures, they would bring you health, wealth, and happiness.

But a Troll's life has become far more complicated. In the '60s these little imps found themselves growing hair in such colors as day-glo pink, green, and orange, and being dressed in outfits resembling aerobics instructors, cheerleaders, or even Playboy bunnies. They had an almost eerie popularity: in the '60s, Trolls were the second-biggest-selling doll behind Barbie, as Troll-mania swept right across America.

Trolls got involved in everyday

life: there was a complete Troll wedding party wearing formal wear. Rhinestone-eyed Trolls called "Groovees" came complete with hipster messages such as "Ban Homework" and "Tell It Like It Is." There were also Santa Trolls and two-headed Trolls.

It is something with an exquisite ugliness that appeals to the kitsch sensibility wrapped up in nostalgia for their most popular era. Trolls come in a variety of sizes, the most common being 3 inches, 5½ inches, and 11 inches. In the '60s, the smallest Trolls were half-inch keyring charms from Hong Kong, and the largest were 18-inch monsters from manufacturers Scandia and Dam Things.

104

✪ ✪

Left: Ramp walkers from the '60s and '70s; just wind 'em up and off they walk. Photos courtesy Hake's Americana, York, PA.

✪ ✪ ✪

Below: Wonderful original '60s "I Dream of Jeannie" bottle containing doll and magic mirror. Photos courtesy Hake's Americana, York, PA.

NO WINDING
PLACE ON INCLINE &
WATCH 'EM WALK
PRINTED IN HONG KONG

✪ ✪

Left: Bullwinkle soaky doll. Soakys were a popular brand of children's bubble bath for children whose bottles were shaped in hommage to popular cartoon figures from the ranks of the animation houses. Photo courtesy Hake's Americana, York, PA.

✪ ✪ ✪

Above: A '70s Tom & Jerry Go Kart with something called (and don't you forget it) Friction Drive. Photo courtesy Hake's Americana, York, PA.

I'm FRED FLINTSTONE
PUSH BUTTON PUPPET
©HANNA-BARBERA PRODUCTIONS INC.
PUSH BUTTON UNDERNEATH BASE
Made in Hong Kong for Kohner Bros. Inc., E. Paterson, N. J.

No. 3991

✪ ✪ ✪

Range of *Flintstones* toys from Pebbles, Barney Rubble, and Bamm-Bamm dolls to Fred & Wilma ramp walker, and a Fred push puppet. As they are all '60s toys, they can prove to be expensive. Photos courtesy Hake's Americana, York, PA.

SOME DAYS **NOTHING** SEEMS TO GO RIGHT!

I'm **YOGI BEAR**
PUSH BUTTON PUPPET
HANNA-BARBERA PRODUCTIONS INC.
PUSH BUTTON UNDERNEATH BASE
Made in Hong Kong for KOHNER BROS. INC., N. Y.
No. 3991

✪ ✪ ✪

Above: Yogi Bear push puppet. Yogi Bear, that lovable beast who was forever feuding with the park ranger was a massively popular cartoon figure with kiddies everywhere. Same era as the *Flintstones*. Photo courtesy Hake's Americana, York, PA.

Top left: Wile E. Coyote figure. Road Runner's relentless tormentor from the hit cartoon serial has been around since the late '40s. Photo courtesy Hake's Americana, York, PA.

✪ ✪

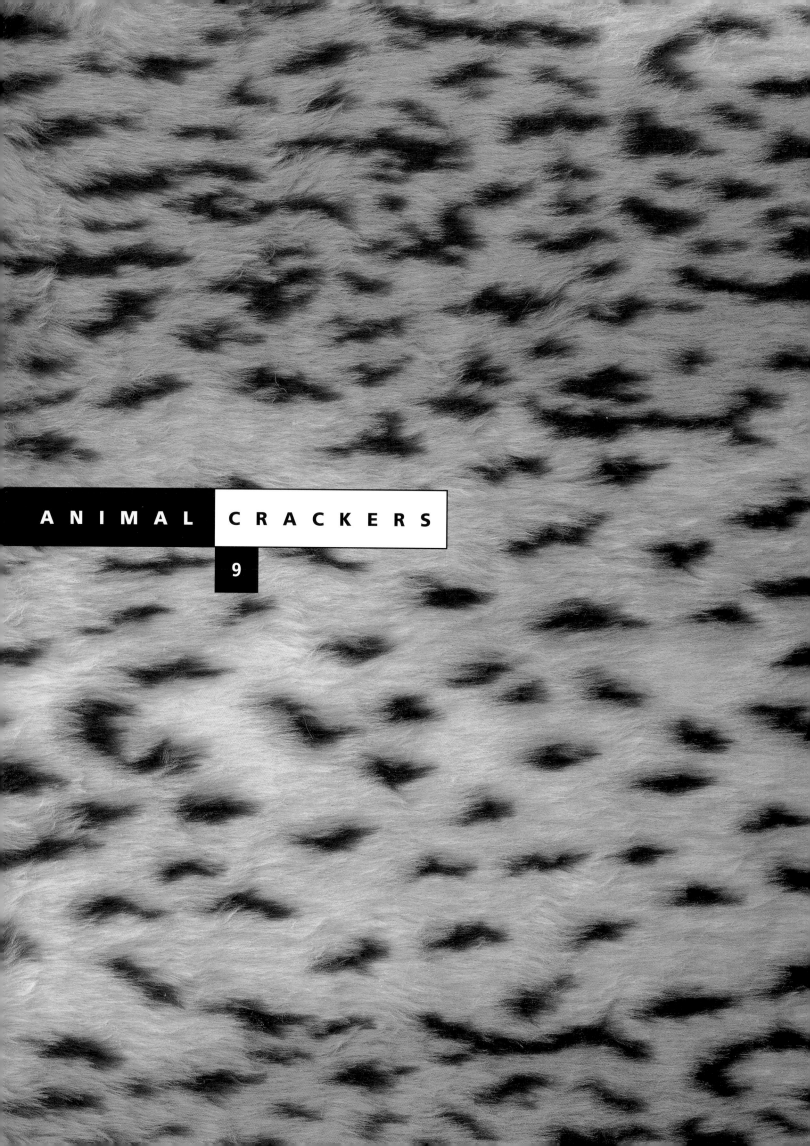

ANIMAL CRACKERS

9

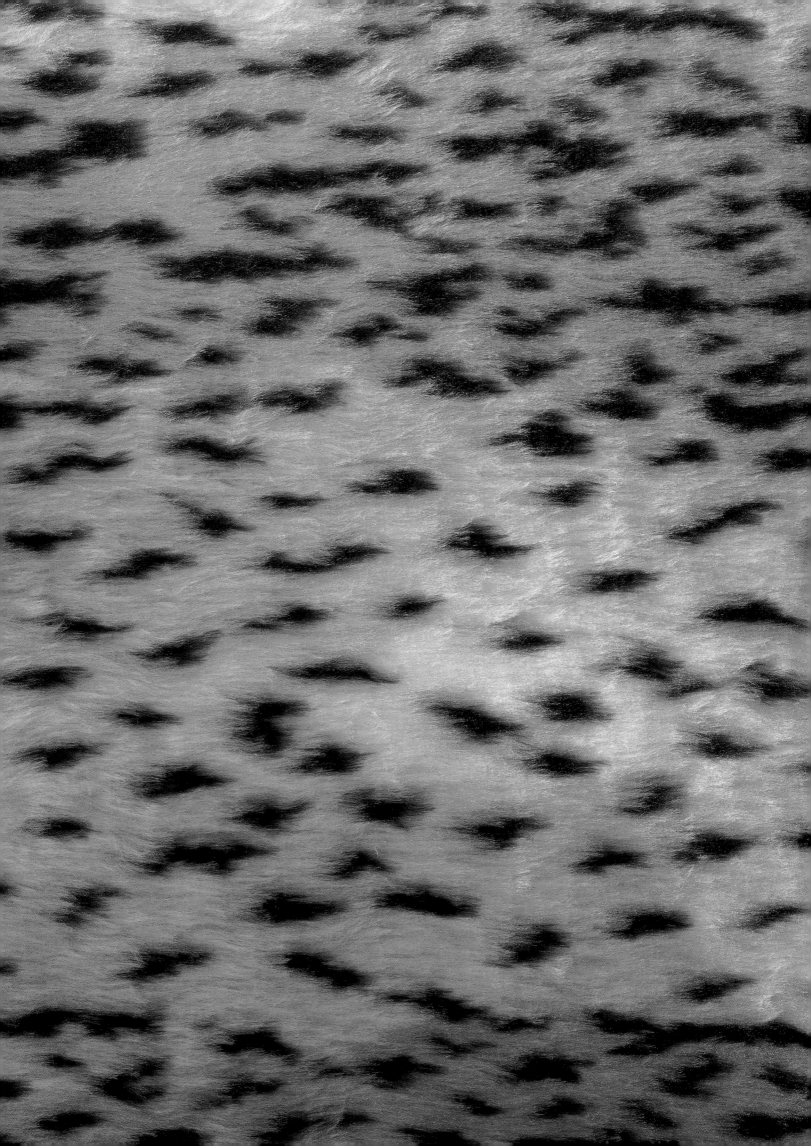

Like Liberace, the poodle is a quintessential symbol of camp kitsch. But the interesting thing is, poodles are not sissies, they aren't even French. They were originally well-respected European gun dogs and they had their coats clipped by hunters as a means of improving their performance. Poodles won fame in the nineteenth century as trick dogs in the French circus. Their extraordinary intelligence and uncanny sense of humor fated them to becoming the clown's clowns, in which their coats were sheared into plumes and ruffles and dyed bright colors to add to the merriment as they jumped through hoops and balanced duckpins on their noses.

In 1960 in America, the poodle became the number-one most popular breed of dog. Remember, this was an era when perspective and scale in all forms of decoration went haywire, so it made perfect sense to take the most decorative of dogs and breed it in tiny sizes guaranteed to be even more amusing than the normal one. In the same kitsch spirit of making cuter things cuter, the poodle's natural colors were supplemented by vegetable dyes that could turn them more shades than nature ever knew. A white-coated dog could be tinted with pastels as pretty as those of a "Coup-de-Ville."

The remarkable thing about poodles is that they once signified chic; they stood for modernity and sophistication. Nowadays, of course, they summon up a memory of a camp comedy of days gone by when a dog was not simply a man's best friend but a multicolored walking, barking topiary.

They crop up quite frequently and are almost certainly authentic '60s produce. For this reason, they make a great starting point for those who wish to begin collecting kitsch plastic animals.

Plastic fun at bathtime with this array of sealife. Plastic fish and sea creatures are relatively cheap and easy to find and, like most plastic animals, look best when displayed in sets.

Left:: A great collection of poodles. A must for every kitsch animal collector. These dogs were popular in the '60s, especially in America.

★ ★ ★

Right: Inflatable safari time!
These air-filled novelties are an
investment for the future
kitsch connoisseur. They tend
to be pricey, but in the years to
come, you may well reap the
rewards financially.

★ ★ ★

Below: Pencil holders were
very popular in the '80s. They
were available in a variety of
shapes, sizes, and colors.

Above: A '60s egg-cup chicks item, a real gem in excellent condition with the extra added value of the original box.

✪ ✪ ✪

Right: Uncle Bulgaria from the '70s British children's TV series "The Wombles" as a child's nightlight.

✪ ✪ ✪

Below: A cheeky cow with movable head. It could be part of a plastic farmyard.

✿ ✿

Left: These animal-themed snow globes date from the '60s to the present day. Collect just a few and you can have your very own tiny kitsch plastic zoo!

✿ ✿

Below: Cute oil-filled ducking birds like these were phenomenally popular on windowsills in the early '70s.

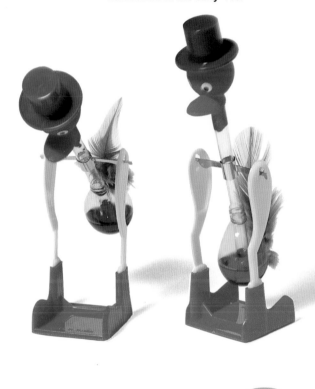

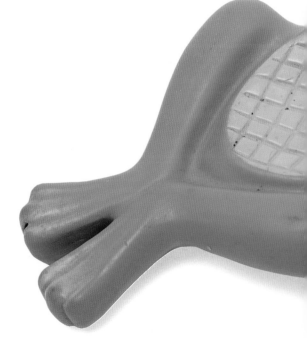

✪ ✪

Left: Popular in the late '70s and early '80s. You can have hours and hours of fun playing with them.

✪ ✪

Right: Originally, filled with candies. Unscrew its middle and the candies tumble out.

✪ ✪

Middle left: A cute yellow piggy money bank. How many coins would fill this piggy?

✪ ✪

Middle Right: A strange cat toy from the '60s looks like a refugee from an Eastern European illustration.

✪ ✪ ✪

Below right: A rare and unusual object from the '60s. Little Robin Redbreast as a toothpick dispenser.

✪ ✪

Below left: A handy frog soapdish. It can float around the bath, so you will never be able to lose your soap.

✪ ✪ ✪

Like the poodle, the nodding dog is a classic kitsch symbol of the '60s. Most popularly sited on a car dashboard, just a light touch on its loose-jointed cranium or a gentle bump in the road will cause its tiny head to bob merrily from side to side.

The first dolls were actually made in China in the mid-1800s from porcelain, bisque, or papier-mâché. But the plastic revolution changed them into cheap, popular fairground prizes and ultimately part of the ancestry of automotive kitsch decor along with furry dice and leopard-skin seat covers. New models are currently being produced.

CELEBRITY CHOICE

10

nd here we arrive at the most expensive part of this guide and one for the true connoisseur. You can pick up some of these dolls for peanuts at rummage sales from some unknowing (and uncaring) vendor. But if you start to collect plastic renditions of the rich and famous, you are very quickly going to know that you have a new hobby.

The type of stuff that has the most value is the spin-off merchandise from TV series and movies. This, if boxed, is a verified and dated objet d'art. The examples that fetch the most are the vintage dolls from the '50s, '60s, and '70s, especially if they were not too common in the first place and if they are modeled on a particular actor, who may have shuffled off the mortal coil, so to speak.

This is where collecting plastic kitsch can actually present you with real investment opportunities. There are many examples here where in a few short years since a TV series or movie finished its run or has had a sequel, the dolls have quadrupled or more their original value. Not even precious metals appreciate in this way. Start now, wise people.

✪ ✪ ✪

Below left: Elvis wiggles those hips in this interesting modern variation on the snow globe, with spinning records instead of snowflakes.

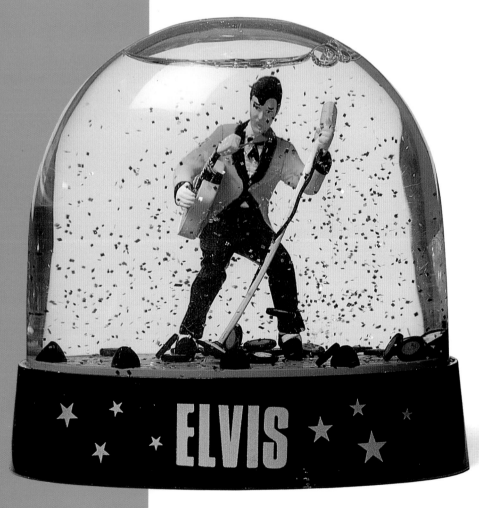

ELVIS

✪ ✪ ✪

Here is a fascinating selection of the King's official and unofficial merchandise. Dolls and watches really are extremely popular, aren't they? Can anyone explain why?

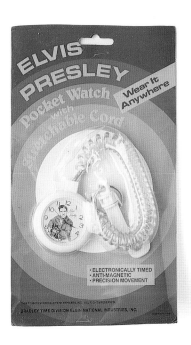

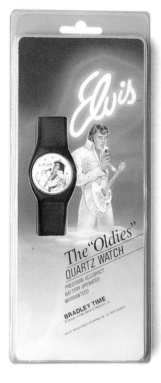

✪✪✪✪

Above right: The Queen and Prince Philip carpet slippers as created by British lampoonists, Spitting Image Productions, in a vinyl-coated latex puppet head and nylon Union Jack fabric. See, plastic after all. Highly collectible.

✪✪✪

Right: Squeaky doll heads of the world superpowers of the late '80s: Thatcher, Reagan, and Gorbachev. Again merchandise from the British TV lampoonists, Spitting Image.

✪ ✪

Left: Marshmallow man from the hit movie, *Ghostbusters*.

✪ ✪

Below: Start investing in your grandchildren's future: Beavis & Butthead are MTV's shmuckster cartoon hits. Modern.

✪ ✪ ✪
Right: Macauley Culkin, the child star from hell? All plastic with (re-)movable limbs.

✪ ✪ ✪
Left: Aaaarghh! There's another one! Macauley Culkin doll with *Home Alone 2* knapsack. Movie tie-in makes him a little more valuable than his twin.

10

The Simpsons, Fox
TV's cartoon hit.

✪ ✪ ✪

Left: Batman and The Penguin. These two dolls hail from different eras: The Penguin dates from the '70s, hence is worth about $50; the modern Keaton/Kilmer-styled Batman is cheaper at $30.

✪ ✪ ✪

Below: Modern *Flintstones* dolls. Fred and Barney modeled on John Goodman and Rick Moranis who played them in the execrable live action movie of 1994.

✪ ✪

Above: Minnie Mouse and *Thunderbird 2* watches.

✪ ✪ ✪ ✪ ✪

Left: This Michael Jackson doll dates from 1984 and in excellent condition can fetch as much as $150!

✪ ✪ ✪

Right: Happy days are here again! This Fonzie doll dates from 1976, authenticated by the Paramount logo. Rare, but has lost his shoes.

☆ ☆ ☆ ☆ ☆

Far left: These "collectors series" Marilyn Monroe dolls come in a variety of outfits and all in these beautiful self-contained display boxes. Here we have Fun Fur Fantasy Marilyn "enveloped in a fanciful creation of glistening gold and fabulous fur" and Spotlight Splendor Marilyn "in the spotlight in a dramatic black gown of pure midnight splendor." Other dolls in this wonderful collection are Silver Sizzle Marilyn and Sparkle Superstar Marilyn.

☆ ☆ ☆ ☆

Left: The Scarecrow from the *Wizard of Oz* remake of 1988 in the highly prized original box.

☆ ☆ ☆ ☆

Below: *Bill & Ted's Excellent Adventure* Wild Stallyn's jam sessions dolls. Just plug in the dolls to the toy amplifiers and turn on the cassette player, and away they rock just like the nerds in the film.

✪ ✪ ✪ ✪

Left: Brenda and Donna character dolls from global hit TV series *Beverley Hills 90210*, now defunct. In the original boxes, each with swimsuits and hairbrushes, they are already worth five times their original price. In just a few years, how much more could they fetch?

✪ ✪ ✪ ✪ ✪

Below: Bionic Man and Woman from the hit TV series date from 1976, General Mills Co.

✪ ✪ ✪ ✪

Above: Donnie out of New Kids On The Block. A very kitsch teenybop group who went the way of all teeny bands: their voices broke, they started shaving, their fans grew out of them and realized they were rubbish.

✪ ✪ ✪ ✪

Right: Speedy Gonzales was the lightening-paced Mexican mouse who has been a popular cartoon figure for nearly fifty years.

★ ★ ★

Above: WWF wrestlers, the
Ultimate Warrior, and an
unidentified Hulk Hogan
lookalike.

★ ★ ★ ★

Left: Worzel Gummidge and
Aunt Sally dolls from the
British children's TV series
of the '70s. With extra
heads!

★ ★ ★ ★ ★

Right: Superheroes together!
Buck Rogers, The Incredible
Hulk, and Superman all date
from the '70s and are very rare.

✪ ✪ ✪ ✪

Left: Dick Tracey and Breathless
(Madonna) dolls from the 1990
movie. Fully clothed and labeled.

✪ ✪ ✪ ✪

Middle left: The Lone Ranger and
Tonto made by Gabriel Industries
in Hong Kong in 1973.
Unfortunately, the Ranger
has lost his hat and both
are *sans chevaux*.

✪ ✪ ✪

Right: Super Mario. Although the
Nintendo Corporation are still
selling this poor old plumber as a
new character in updates of the
classic games and other
merchandising spin-offs, he is
their Mickey Mouse. A valuable
investment.

✪✪✪✪

Left: Slater from the teenie TV series *Saved By The Bell*. Tiger Toys 1995. Another one for the future.

✪✪✪✪

Below: Anne of Green Gables made by Ideal Toys in 1975. Wonderfully rare and fully dressed.

✪✪✪✪

Bottom: *The A-Team*'s Mr. T dolls from the early '80s. The originals came with different wardrobes full of clothes.

✪ ✪ ✪ ✪ ✪ ✪ ✪ ✪ ✪ ✪
✪ ✪ ✪ ✪ ✪ ✪ ✪ ✪ ✪ ✪
✪ ✪ ✪ ✪ ✪ ✪ ✪ ✪ ✪ ✪
✪ ✪ ✪ ✪!

Above: Original mid-'60s Beatle dolls in ultra-good condition. These are almost priceless. Approximate value $1,000 as a set! Photos courtesy Hake's Americana, York, PA.

✪ ✪ ✪

Left: Captain Scarlet from Gerry Anderson's '60s British TV puppet series.

✪ ✪ ✪

Center right: A very handy pocketbook and purse for "mad" money, courtesy Laverne & Shirley, those '70s TV hicksters. Photo courtesy Hake's Americana, York, PA.

✪ ✪ ✪

Far right: Babe Ruth statuette.

ANYTHING GOES

11

Almost as interesting as the mind of the obsessive collector (and almost as much an outsider of society) must be the designers and entrepreneurs behind much of the produce that is featured in this book. What could have been going on in the minds of the individuals who conceived, planned, wrote manufacturers' specifications, produced, and marketed such goods? What sort of warped or inspired vision did the entrepreneur have who made a success of the nodding dogs in chapter 9. Did this person take a prototype and technical drawings and specifications to the patent office? Were they able to obtain a business-development loan for such a transparently madcap scheme from their local bank loan officer? Did they become millionaires? The answers to the last three of these questions is yes, the others we can only speculate upon.

✪

Below left: Christmas comes but once a year and with it come presents, sleigh bells, snow, Christmas trees, holly, mistletoe, and snowmen, in plastic. With die-hard collectors leading the charge for antique Christmas collectibles, such as hand-blown decorations, prices have sky-rocketed, but there are hordes of collectors after plastic decorations because they are fun, low-priced and kitschy.

Just a few years ago, these plastic items were considered to be junk, now there is a whole new generation of collectors who remember the plastic decorations of Mom and Dad's or Grandma and Grandpa's tree.

**Above: Contemporary
household objects with novelty
twists such as this see-through
telephone, serve not only to
deconstruct the communications
industry before your eyes but,
when lined up side by side with
a psychedelic watch from the
Swiss "pop" watch firm, Swatch,
aim themselves very firmly at
the kitsch-conscious consumer. A
long-term investment.**

**Below right: The Rubik's Cube
was a runaway success as a
novelty puzzle in the '80s. Many
variations began to be produced
from the original cubed formula.
Its vivid, almost Mondrian-esque
primary colors are a symbol of a
confident and optimistic
capitalistic era. This seems to
date them as objects from a time
gone past as much as any kidney
shape from the '50s or op-art
pattern from the '60s. Kitsch
status to be bestowed upon
these very soon.**

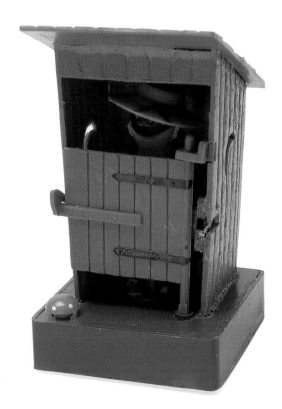

✪ ✪ ✪

Below: A gaggle of spherical and circular plastic objects from several different decades. With the exception of the blue lightshade at the back, they are all "executive" toys of one description or another. Toys for adults to provide a distraction from the stresses of ordinary life and work: a modular molecular puzzle, a chrome-look plastic ball tree, and a mine-shaped ball.

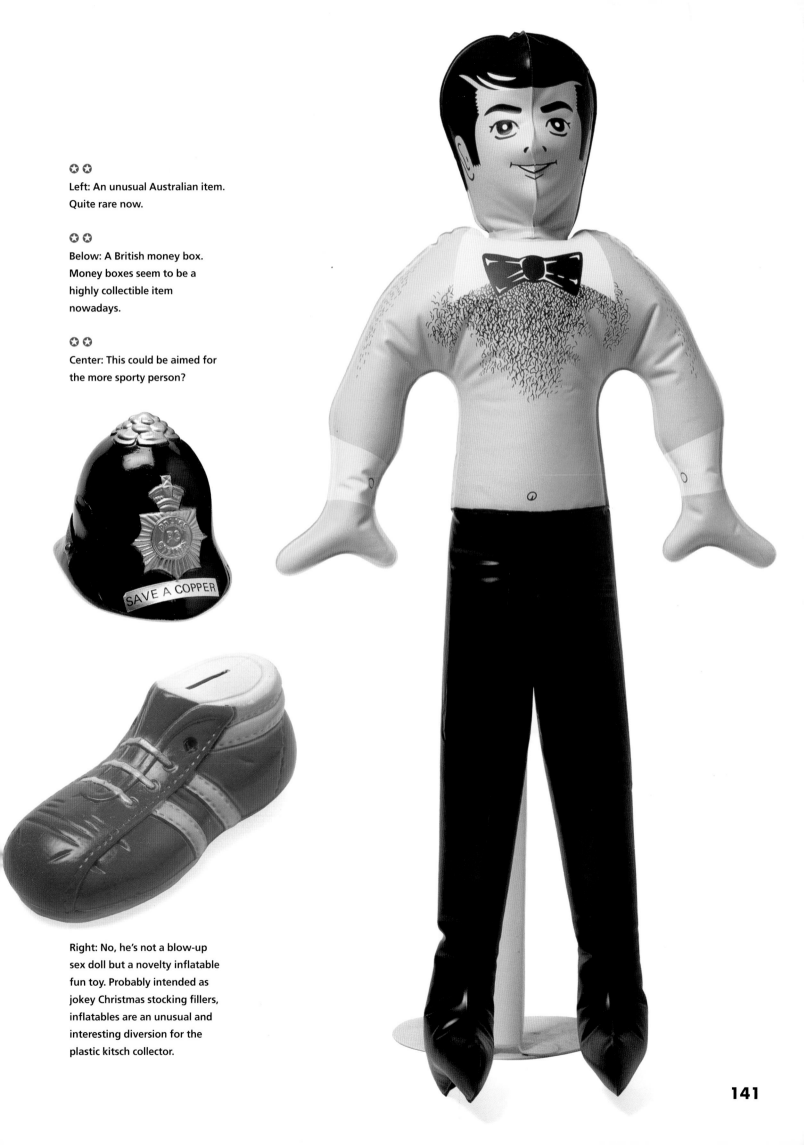

✪ ✪
Left: An unusual Australian item. Quite rare now.

✪ ✪
Below: A British money box. Money boxes seem to be a highly collectible item nowadays.

✪ ✪
Center: This could be aimed for the more sporty person?

SAVE A COPPER

Right: No, he's not a blow-up sex doll but a novelty inflatable fun toy. Probably intended as jokey Christmas stocking fillers, inflatables are an unusual and interesting diversion for the plastic kitsch collector.

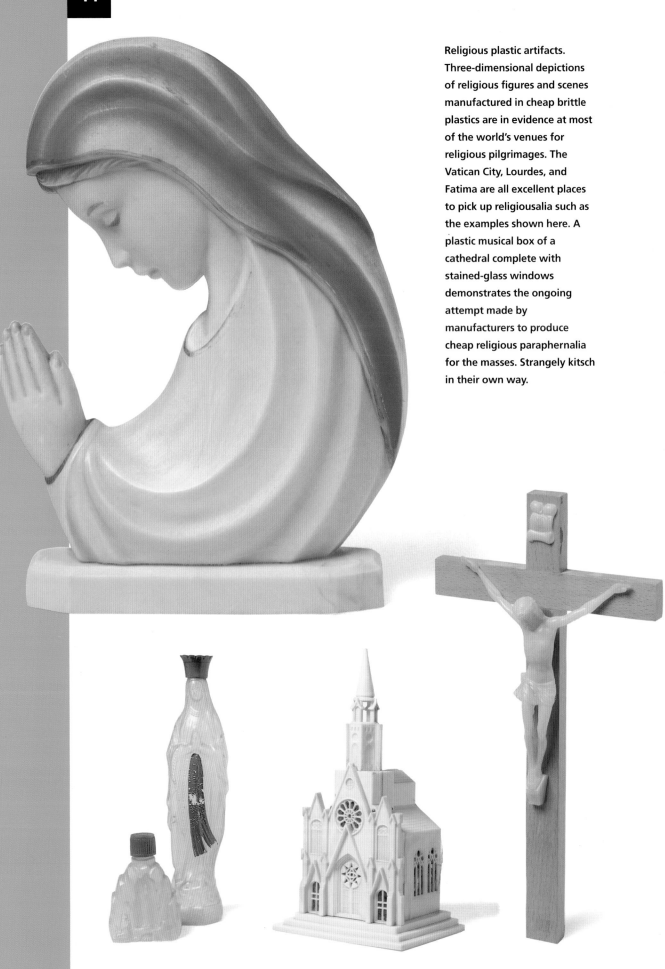

Religious plastic artifacts. Three-dimensional depictions of religious figures and scenes manufactured in cheap brittle plastics are in evidence at most of the world's venues for religious pilgrimages. The Vatican City, Lourdes, and Fatima are all excellent places to pick up religiousalia such as the examples shown here. A plastic musical box of a cathedral complete with stained-glass windows demonstrates the ongoing attempt made by manufacturers to produce cheap religious paraphernalia for the masses. Strangely kitsch in their own way.

✪ ✪ ✪

Top: Unusual and interesting
outside porch light. The roses
look real. Quite difficult to find.

✪ ✪

Below right: Another money
box. This time in the form of a
policeman.

While every effort has been made to trace and acknowledge all copyright holders, we would like to apologise should any omissions have been made.

Many thanks to the collectors who provided their prized specimens for this tome; Carolyn and James and their fabulous kitsch retail enterprise Flying Duck Enterprises of Creek Rd, Greenwich, London (0181-858-1964), Steve Rumney who has scoured the world to add to his vast exhibition of objets d'art which can be viewed at his internet website archive and library resource at the following address: http//www.1off.com, Kirsten Hardie and Bournemouth & Poole College of Art & Design, Ted Hake Collectibles, Toy Scouts Inc., Jeremy Thomas for his excellent photography, Percy Reboul for making the introduction in the first place and my wife, Brenda and daughter Kathleen for their undying love, support and understanding over my obsession with kitsch.

PICTURE CREDITS

t = top, c = center, b = bottom

pages 90(t), 91(b), 106(b), 107(b) "This book makes reference to various Disney copyrighted characters, trademarks, marks and registered marks owned by Disney Enterprises, Inc."

pages 88(c) "Reproduced with the permission of Esso UK plc."

pages 6(b), 85(t), 93(c), 103(b), 104(b), 105(b), 135(t), 136(c) "Photos credited to Hake's Americana are of items that appeared in their bi-monthly mail and telephone bid auctions. Each auction offers over 3,000 items and a free sample copy is available by writing Hake's Americana Dept. 352, P.O. Box 1444, York, PA 17405 USA."

pages 106(b), 107(t0, 126(c), 127(b) reproduced with the permission of Hanna Barbera Inc.

pages 86, 87 reproduced with the permission of "McDonald's Restaurants Ltd."

page 142 reproduced with the permission of Spitting Image Productions.

pages 94, 95 reproduced with the permission of "Toy Scouts, Inc."

pages 105(t), 106(c), 107(t), 126(t), 136(b), 137(b) reproduced with permission of Twentieth Century Fox.

US SOCIETIES

National Fantasy Club for Disneyana Collectors
PO Box 19212, Irvine,
CA 92713

Toy Scouts, Inc.,
137 Casterton Ave.
Alcron, OH 44303 USA
tel: 330-836-0668/fax: 330-869-8668
email: toyscout@newreach.net website: http://www.comonline.com/toyscouts.
Toy Scouts serve a large, international clientele with their quarterly sale and auction catalogs each year. President Bill Bruegman shares his insights on toys and the art of collecting in his price guides.

UK SOCIETIES

The UK Disney Club
31 Rowan Way, Exwick
Exeter EX4 2DT

Plastic Historical Society
2 Park Avenue
Radlett, Herts WD7 7EA
(Please send a SAE.)